Ecological Risks Associated with the Destruction of Chemical Weapons

NATO Security through Science Series

This Series presents the results of scientific meetings supported under the NA Programme for Security through Science (STS).

Meetings supported by the NATO STS Programme are in security-related priority areas of Defence Against Terrorism or Countering Other Threats to Security. The types of meeting supported are generally "Advanced Study Institutes" and "Advanced Research Workshops". The NATO STS Series collects together the results of these meetings meetings are co-organized by scientists from NATO countries and scientists from NA "Partner" or "Mediterranean Dialogue" countries. The observations and recommendations made at the meetings, as well as the contents of the volumes in the Series, reflect those of participants and contributors only; they should not necessarily be regarded as reflecting NATO views or policy.

Advanced Study Institutes (ASI) are high-level tutorial courses to convey the latest developments in a subject to an advanced-level audience

Advanced Research Workshops (ARW) are expert meetings where an intense b informal exchange of views at the frontiers of a subject aims at identifying directions f future action

Following a transformation of the programme in 2004 the Series has been re-named and re-organised. Recent volumes on topics not related to security, which result from meetings supported under the programme earlier, may be found in the NATO Science Series.

The Series is published by IOS Press, Amsterdam, and Springer, Dordrecht, in conjunction with the NATO Public Diplomacy Division.

Sub-Series

A. Chemistry and Biology	Springer
B. Physics and Biophysics	Springer
C. Environmental Security	Springer
D. Information and Communication Security	IOS Press
E. Human and Societal Dynamics	IOS Press

http://www.nato.int/science
http://www.springer.com
http://www.iospress.nl

Series C: Environmental Security

Ecological Risks Associated with the Destruction of Chemical Weapons

edited by

Vladimir M. Kolodkin
Institute of Natural and Technogenic Disasters,
Udmurt State University,
Izhevsk, Russia

and

Wolfgang Ruck
Institute of Ecology and Environmental Chemistry,
University of Lüneburg,
Germany

Springer

Proceedings of the NATO Advanced Research Workshop on
Ecological Risks Associated with the Destruction of Chemical Weapons
Lüneburg, Germany
22–26 October 2003

A C.I.P. Catalogue record for this book is available from the Library of Congress.

ISBN-10 1-4020-3136-X (PB)
ISBN-13 978-1-4020-3136-6 (PB)
ISBN-10 1-4020-3135-1 (HB)
ISBN-13 978-1-4020-3135-9 (HB)
ISBN-10 1-4020-3137-8 (e-book)
ISBN-13 978-1-4020-3137-3 (e-book)

Published by Springer,
P.O. Box 17, 3300 AA Dordrecht, The Netherlands.

www.springer.com

Printed on acid-free paper

Printed in the Netherlands.

Contents

Contributing Authors

Name	Affiliation	Contact
Arbusov, Gennadiy	Federal State Unitary Establishment Russian Scientific Center Applied Chemistry, Saint-Petersburg (Russia)	arseny@mail.lanck.net
Baljasnicov, I. A.	The Bryansk regional authority of natural resources and environmental protection, Bryansk (Russia)	
Benning, Helmut	German Institute of Rubber Technology, Eupener Straße 33, 30519 Hannover (Germany)	voice +49 511 84201-47 Helmut.Benning@DIKautschuk.de
Birke, Volker	University of Applied Sciences-North-East Lower Saxony, Suderburg, Germany, Faculty of Civil Engineering (Water and Environmental Management), Steinweg 4, 30989 Gehrden (Germany)	voice +49 5108 9217-30 fax +49 5108 9217-39 birke@fhnon.de
Bouchmakina, Ioulia	The Centre for State Sanitary and Epidemiological Inspection in Udmurt Republic, Izhevsk (Russia)	gsenr@udmnet.ru
Bykov, A. A.	SRC «Ecosafety», Moscow (Russia)	

Carlsen, Lars	Tacis Chemical Weapons III Project, Novocheboksarsk (Russia) Awareness Center, Hyldeholm 4, DK-4000 Roskilde (Denmark)	LC@AwarenessCenter.dk
Churakov, Alexander	The Izhevsk State Medical Academy, 199 Revolutsionnaya, 426033 Izhevsk (Russia)	gsenr@udmnet.ru
De Bisschop, Herbert C.	Royal Military Academy, Department of Chemistry, 30 Avenue Renaissance, B 1000 BRUSSELS (Belgium)	herbert.de.bisschop@rma.ac.be
Decker, John A.	Department of Health and Human Services, Centers for Disease Control and Prevention, 30333 Atlanta, Georgia (U.S.A)	jad4@cdc.gov
Dobrylko, Irina	I.M.Sechenov Institute of Evolutionary Physiology and Biochemistry. St.Petersburg (Russia)	
Egorov, Alexander	D.I. Mendeleyev Russian Chemical-Technological University,, Minsskaya 9, 125047 Moscow (Russia)	egorov@muctr.edu.ru
Ermolayeva, Elena	Research Institute of Hygiene, Occupational Pathology and Human Ecology, p/o Kugmolovsky, 188663 Leningradsky Region, St.Petersburg (Russia)	gipeh@lens.spb.ru gipeh@spb.cityline.ru
Fedotova, Irina	Udmurtia State University, 1 Universitetskaya, 426034 Izhevsk (Russia)	intd@wing.uni.udm.ru
Fendler, Roland	Federal Environment Agency *(Umweltbundesamt)*, P.O. Box 330022, 14191 Berlin (Germany)	Roland.Fendler@uba.de
Fominykh, Anatoly A.	Chairman of the Committee on Conventional Problems of Chemical Weapons under the Government of the Udmurt Republic, 220 Pushkinskaya, 426008 Izhevsk (Russia)	himcom@udmnet.ru

Garbotz, Christine	Chair of Chemical Engineering and Hazardous Wastes, Brandenburg University of Technology, Cottbus (Germany)	
Glashkina, Lidia	Research Institute of Hygiene, Occupational Pathology and Human Ecology, p/o Kugmolovsky, 188663 Leningradsky Region, St.Petersburg (Russia)	gipeh@lens.spb.ru gipeh@spb.cityline.ru
Glazun, I. N.	Bryansk State Academy of Engineering and Technology, Bryansk (Russia)	
Goncharov, Nikolay	Research Institute of Hygiene, Occupational Pathology and Human Ecology, p/o Kugmolovsky, 188663 Leningradsky Region, St.Petersburg (Russia)	gipeh@lens.spb.ru gipeh@spb.cityline.ru
Gorsky, Vladimir G.	Russian Research Center (GOSNIIOKHT), 3 Proezd Perova Polya 5 A – 6, 111141 Moscow (Russia)	
Hill, Herbert H.	Department of Chemistry, Washington State University Pullman, Washington 99164-4630 (U.S.A.)	hhhill@wsu.edu
Ionov, Leonid	Udmurtia State University, Izhevsk , 1 Universitetskaya, 426034 Izhevsk (Russia)	
Ivanov, V. P.	Bryansk State Academy of Engineering and Technology, pr. St. Dimitrova 3, 241037 Bryansk (Russia)	Info@ivp.bryansk.ru
Kolodkin, Vladimir Mikhailovitsh	Institute of Natural and Technogenic Disasters, Udmurt State University, 1 Universitetskaya, 426034 Izhevsk (Russia)	kolodkin@ wing.uni.udm.ru
Korepanov, Mikhail A.	Institute of Applied Mechanics, Ural Branch Russian Academy of Sciences, Izhevsk (Russia)	
Krutzsch, Walter	Deputy Head of Delegation of the GDR to the CWC Negotiations, Löwestrasse 22, 10249 Berlin	Walter.Krutzsch@ t-online.de

Kudankina, Tatyana G.	Center for State Sanitary Supervision No 29, Novocheboksarsk (Russia)	cgsen29@cbx.ru
Kurguzkin, Mikhail	Ministry of Natural Resources and Environmental Protection of the Udmurt Republic, 1 Sovetskaya Street, 426051 Izhevsk (Russia)	mpr@udmnet.ru
Kuznetsov, Anatoly	Research Institute of Hygiene, Occupational Pathology and Human Ecology, p/o Kugmolovsky, 188663 Leningradsky Region, St.Petersburg (Russia)	gipeh@lens.spb.ru gipeh@spb.cityline.ru
Kuznetsov, Sergey	I.M.Sechenov Institute of Evolutionary Physiology and Biochemistry. St.Petersburg (Russia)	
Langard, Sverre	The Centre for Occupational and Environmental Medicine, The National Hospital, Oslo (Norway)	
Laskin, Boris	Federal State Unitary Establishment Russian Scientific Center Applied Chemistry, Saint-Petersburg (Russia)	
Lipanov, Alexey M.	Institute of Applied Mechanics, Ural Branch Russian Academy of Sciences, 34 T. Boramzina Str., 426067 Izhevsk (Russia)	elle@udman.ru
Lystsov, V. N.	RRC «Êurchatov Institute», 1 Kurchatov Pl., 123182 Moscow (Russia)	vitalil@pike.pike.ru
Makarova, Lyudmila	Udmurtia State University, Izhevsk (Russia)	
Matousek, Jiri	EU Research Centre of Excellence for Environmental Chemistry and Ecotoxicology, Faculty of Science, Masaryk University Brno, Kamenice 3, CZ-625 00 Brno (Czech Republic)	matousek@ recetox.muni.cz

Mattik, Jörg	University of Applied Sciences-NE Lower Saxony, Suderburg, Germany, Department of Civil Engineering (Water and Environmental Management), Steinweg 4, 30989 Gehrden (Germany)	
Meert, Christel	Royal Military Academy, Department of Chemistry, 30 Avenue Renaissance, B 1000 BRUSSELS (Belgium)	
Michailov, Anatoliy	Russian Federal Nuclear Center. All-Russian Research Institute of Experimental Physics, (RFNC-VNIIEF), Sarov (Russia)	
Mindukshev, Igor	I.M.Sechenov Institute of Evolutionary Physiology and Biochemistry. St.Petersburg (Russia)	
Motkin, Gennadiy	Head of Laboratory in the Institute for Market Economy Problems of Russian Academy of Sciences (IPR RAN), 47 Nahimovsky prospect, 117418, Moscow (Russia)	motkinga@mtu-net.ru (095) 332-4224 (work/fax) (095) 935-4792 (home)
Murin, Aleksey	Research Institute of Natural and Technogenic Disasters, Udmurtia State University, 1 Universitetskaya, 426034 Izhevsk (Russia)	
Murzin, N. V.	RRC «Êurchatov Institute», Moscow (Russia)	
Nagorny, Sergey	Research Institute of Hygiene, Occupational Pathology and Human Ecology, p/o Kugmolovsky, 188663 Leningradsky Region, St.Petersburg (Russia)	gipeh@lens.spb.ru gipeh@spb.cityline.ru
Nartov, D. I.	Bryansk State Academy of Engineering and Technology, Bryansk (Russia)	
Padayhag, Joseph	Department of Health and Human Services, Centers for Disease Control and Prevention (CDC), 30333 Atlanta, Georgia (U.S.A.)	ozp9@cdc.gov JPadayhag@cdc.gov

Pavlikova, Nina P.	Analytical Laboratory of Shop 83, Detached Plant No 4 of JSC Khimprom, Novocheboksarsk (Russia)	
Perneel, Christiaan P.	Royal Military Academy, Department of Mathematics, 30 Avenue Renaissance, B 1000 BRUSSELS (Belgium)	
Petrin, Stanislav	Russian Federal Nuclear Center. All-Russian Research Institute of Experimental Physics, (RFNC-VNIIEF), Mira St. 37, 607188 Sarov (Russia)	root@gdd.vniief.ru v_tiunov@mail.ru
Petrina, Lada	Russian Federal Nuclear Center. All-Russian Research Institute of Experimental Physics, (RFNC-VNIIEF), Sarov (Russia)	
Petrov, Vadim	Institute of Applied Mechanics UB RAS, 34 T. Boramzina Str., 426067 Izhevsk (Russia)	
Radilov, Andrey	Research Institute of Hygiene, Occupational Pathology and Human Ecology, p/o Kugmolovsky, 188663 Leningradsky Region, St.Petersburg (Russia)	gipeh@lens.spb.ru gipeh@spb.cityline.ru
Reshetnikov, Sergey	Udmurtia State University, 1 Universitetskaya, 426034 Izhevsk (Russia)	kolodkin@ wing.uni.udm.ru
Rogers, Harvey W.	Department of Health and Human Services, Centers for Disease Control and Prevention, Atlanta, Georgia (U.S.A)	
Ruck, Wolfgang	Institute of Ecology and Environmental Chemistry, University of Lüneburg (Germany)	ruck@uni-lueneburg.de
Runne, Dietlind	University of Applied Sciences-NE Lower Saxony, Suderburg, Germany, Department of Civil Engineering (Water and Environmental Management), Steinweg 4, 30989 Gehrden (Germany)	

Saveljeva, Elena	Research Institute of Hygiene, Occupational Pathology and Human Ecology, p/o Kugmolovsky, 188663 Leningradsky Region, St.Petersburg (Russia)	gipeh@lens.spb.ru gipeh@spb.cityline.ru
Savitskaya, Tatiana	D.I. Mendeleyev Russian Chemical-Technological University, Moscow (Russia)	
Sheluho, V. P.	Bryansk State Academy of Engineering and Technology, Bryansk (Russia)	
Shkayeva, Irina	Research Institute of Hygiene, Occupational Pathology and Human Ecology, p/o Kugmolovsky, 188663 Leningradsky Region, St.Petersburg (Russia)	gipeh@lens.spb.ru gipeh@spb.cityline.ru
Shvetsova-Shilovskaya, Tatjana	State Research Institute of Organic Chemistry and Technology (SRIOCT), Novogireevskaya 44/28-327, 111397 Moscow (Russia)	shvezova@cityline.ru in55@narod.ru
Silberschmidt, Martin	Tacis Chemical Weapons III Project, Novocheboksarsk (Russia)	
Smedts, Bart R.	Royal Military Academy, Department of Chemistry, 30 Avenue Renaissance, B 1000 BRUSSELS (Belgium)	
Spyra, Wolfgang	Brandenburg Technical University of Cottbus, Chair of Chemical Engineering and Hazardous Wastes, P.O. Box 10 13 44, 03013 Cottbus (Germany)	wolfgang.spyra@tu-cottbus.de
Steiner, Wes E.	Department of Chemistry, Washington State University Pullman, Washington 99164-4630 (U.S.A.)	
Stock, Thomas	Tacis Chemical Weapons III Project, Novocheboksarsk (Russia); Dynasafe, Düsseldorfer Str. 138, 45481 Mülheim (Germany)	thomas.stock@dynasafe.de

Trubachev, Aleksey	Institute of Applied Mechanics UB RAS, Izhevsk (Russia)	
Tsibulskaya, Elena	Research Institute of Hygiene, Occupational Pathology and Human Ecology, p/o Kugmolovsky, 188663 Leningradsky Region, St.Petersburg (Russia)	gipeh@lens.spb.ru gipeh@spb.cityline.ru
Tsimbal, Fyodor	Research Institute of Hygiene, Occupational Pathology and Human Ecology, p/o Kugmolovsky, 188663 Leningradsky Region, St.Petersburg (Russia)	gipeh@lens.spb.ru gipeh@spb.cityline.ru
Tukhvatullin, Zufar A.	State Corporation «Votkinskiy zavod», Votkinsk (Russia)	
Vasilyev, Leonid	Research Institute of Natural and Technogenic Disasters, Udmurtia State University, 1 Universitetskaya, 426034 Izhevsk (Russia)	
Vollerthun, Tina	Brandenburg Technical University of Cottbus, Chair of Chemical Engineering and Hazardous Wastes, P.O. Box 10 13 44, 03013 Cottbus (Germany)	tina.vollerthun@tu-cottbus.de
Winkelmann, Kay	Brandenburg Technical University of Cottbus, Chair of Chemical Engineering and Hazardous Wastes, P.O. Box 10 13 44, 03013 Cottbus (Germany)	kay.winkelmann@tu-cottbus.de
Zabrodine, Nicolay	The Centre for State Sanitary and Epidemiological Inspection in Udmurt Republic, 106 Lenina Str., 426009 Izhevsk (Russia)	
Zenkevich, Igor	Chemical Research Institute of St.-Petersburg State University, St.Petersburg (Russia)	
Zlatovic, Dragan	L+Z Entsorgungsdienste für Starkstromanlagen GmbH, Am Bubenpfad 2, 67065 Ludwigshafen am Rhein (Germany)	voice +49 621 5793-191 zlatovic.d@luz-trafo.de
Zubtsovsky, Nikolay	Udmurtia State University, Izhevsk (Russia)	

Welcome Address by the Federal Minister of Defence, Dr. Peter Struck

I wish you all a warm welcome. I greatly appreciate the fact that so many national and international delegates are here today for the opening of this major workshop. It gives you the opportunity for an in-depth exchange of your experiences with one of the most difficult issues of disarmament – the speedy and environmentally sound destruction of chemical weapons.

Today, the international community is called upon in a host of ways to increase global security – by fighting international terrorism, by managing crises and conflicts in many places around the world, in combating the proliferation of weapons of mass destruction.

The variety of the threats requires us to make comprehensive, foresighted and common provisions for security. NATO and the EU, together with partners like Russia, for instance, have a central role to play in this.

The aim of the policy is to eliminate military threat potentials and eradicate their political and social causes as well as to build lasting stability. This can only be achieved by adopting a multidimensional approach. Military means alone are not enough, but they cannot be dispensed with either.

This multidimensional approach is reflected in the current crisis management operations in the Balkans or in Afghanistan. We are endeavouring to come up with overall political concepts. We are providing support along the hard road to democracy, considerable shares of international security forces and well-trained police officers. We are assisting in the setting up of a functioning administration. And and we are improving the prospects for economic development.

The issues of international crisis management, however, should not let us forget other security issues, which too easily escape our notice in the slipstream, as it were, of topical debate. The implementing of disarmament commitments is one of them.

We would jeopardise the tremendous successes yielded in conventional, nuclear and chemical disarmament by the overcoming of the East-West conflict if we were not willing to work together to solve the practical problems connected with their implementation. Of all the areas of cooperation, it is in the destruction of weapons of mass destruction and their delivery systems that has illustrated the epochal political change in the last ten years or so in an impressive way. This cooperation increases our common security, reduces the risks of proliferation and builds confidence.

The destruction of Russia's stocks of around 40,000 tons of chemical agents is a tremendous task, which cannot be performed without international support. I am glad that Germany is also able to make a substantial contribution towards it by providing assistance for the destruction plant in Gorny and the second one in Kambarka. The destruction of this Cold War legacy can now begin.

This effort will take many years and only yield the desired success if numerous technical problems are solved and ecological risks are dealt with satisfactorily. In this context, your visit to GEKA – the federal company responsible for the disposal of chemical agents and discarded arms and ammunition – in Munster will certainly be useful.

Only if action is taken to ensure that chemical agents are destroyed in an economically sound manner can people living in the vicinity of the destruction plants be convinced that there is no serious alternative to complete and controlled disposal.

This workshop, financed by NATO and the Russian Federation and organised by the University of Lüneburg, is an important step towards accomplishing this complex task.

I wish both the organisers and the participants every success with the conference.

Thank you.

Foreword

Prof. Dr. Vladimir Mikhailovitsh Kolodkin[1], Prof. Dr.-Ing. Wolfgang Ruck[2]
[1]*Institute of Natural and Technogenic Disasters, Udmurt State University, Izhevsk (Russia),*
[2]*Institute of Ecology and Environmental Chemistry, University Lüneburg (Germany)*

During the Cold War a whole arsenal of deadly chemical weapons was allowed to build up on both sides of the ideological divide. Happily, today the problems are reversed. Expertise is now required in the field of safe and environment-friendly disposal of chemical weapons and cleaning up of contaminated sites all around the world, but not least in the ex-Soviet-led countries.

The participants and speakers to the NATO-Russia advanced research workshop on the "Ecological Risks Associated with the Destruction of Chemical Weapons", hosted by the University of Lüneburg on 22nd - 26th October, 2003, therefore, came from many different parts of the world.

Of the eight countries represented at the workshop, two were ex-Eastern-Block, and six were Western countries. Yet the West was by no means overrepresented. On the contrary, the Russian expert-speaker contingent, with 33 participants, did justice to the size of their country – and to their chemical-weapons problem – and provided the majority of active participants. In all, there were 57 participants, of which 11 dispatched from the TACIS project "The development of the chemical weapons" facility at the detached plant No 4 of OAO Khimprom, Novocheboksarsk.

The main concern of this four-day workshop, and the chief occupation of the participants, namely the elimination of existing Chemical Weapons

V.M. Kolodkin and W. Ruck (eds.),
Ecological Risks Associated with the Destruction of Chemical Weapons, 1–4.

(CW), is clearly stated in the Convention on the Prohibition of the Development, Production, Stockpiling and Use of Chemical Weapons and on Their Destruction. The Convention, which was the subject of many discussions, requires every Signatory to destroy not only all chemical weapons, but also all chemical weapons production facilities on its territory, as well as any chemical weapons it may have left behind on the territory of another Signatory.

To the lay person, lethal weapons' destruction sounds like a daunting task. Yet, the speakers made the environment-friendly disposal of chemical weapons look like mere routine.

Throughout these highly productive four days, next to the conferences, a lot of discussions took place which gave the participants the opportunity to exchange their own experiences and their views, but also to ask the questions they were burning to ask.

Apart from the ethics, political and international-law issues associated with chemical weapons and their destruction, there is a wide range of more concrete problems related to the real disposal operations for this type of warfare material. These concrete issues provided the quasi totality of the topics for the workshop discussions. The majority of the workshop contributions, all of which were in the English language, dealt with the environmental problems linked with the destruction of chemical-warfare agents (CWA). These included the quantitative assessments of ecological risks posed by CW destruction technologies and facilities, the influence on natural resources and ecological systems as well as the modelling of ecological reactions resulting from chemical and physical properties of CWA.

Toxicity, health problems and the human cost of the polluted-environment legacy provided material for many other workshop topics, some titles sounded somehow rather medical, for example "New understanding on pathogenesis of delayed effects of RVX low-dose chronic exposure" . Monitoring and detection, including the review of existing workplace CWA monitoring systems as well as the review of existing public health CWA information and monitoring systems was offered as a solution to these problems.

Purely chemistry-related contributions, such as the lessons learned while selecting and using CWA trace detection systems, e.g. lewisite detection offered another batch of contributions.

Some papers concentrated on the Convention itself, namely on its shape after the Review Conference. Environmental-law issues such as national laws and regulations for risk prevention related to CWA destruction were also tackled.

A few contributions dealt with administrative issues, others were concerned by insurance matters, a point that should not be forgotten, in view of all the dangers and property-owner liability associated with these health- and life-threatening activities.

This conference wrote safety issues with a capital "S", as special emphasis was put on purely technical risk as well as the risks posed by terrorism. A very topical issue right now! The main points discussed included safety aspects associated with the selection of CWA destruction technologies, emergency planning and disaster preparedness, as well as national limit values.

Another risk is seen in the lack of appreciation from the population, or from the administration, for this huge and complicated task. Who wants to live near a CW facility, who wants to pay taxes to support the disarmament process, who knows the real financial demands of chemical weapons' destruction? A non-informed public will be a non-committed public, and will be reluctant to push their parliament to pass legislation for financing the disposal of these weapons, thereby putting chemical-warfare agents' disposal at risk. There is, therefore, still a lot to be done in the areas of crisis management, public information procedures and information systems and lessons learned in implementing public outreach regulations must be drawn.

Though the workshop mainly concentrated on concrete issues related to destruction of chemical warfare agents and manufacturing operations thereof, the political nature of the conclusions reflect the difficult intricacies surrounding these horrifying weapons.

A fundamental point, which the environmental discussion outshined, is that all weapons are, and NATO is, by definition, a military issue. It comes as no surprise, therefore, that military orders or withholding of permission accounted for a considerable number of cancellations from NATO countries, as some would-be contributors did not receive approval for their papers from their superiors. This brings us to another one of the discussions namely that the secrecy, which naturally surrounds all military operations, often impedes proper scientific interchange and renders open discussion difficult.

The contributions made by these many enthusiastic as well as highly qualified speakers, some of whom travelled from as far as Japan or the USA, not to mention the farthest depths of the Russian Federation, are compiled in this book. They offer the reader very clear, manifold and above all expert views of the intricate anatomy of the task facing chemists dedicated to safety and to the preservation of the environment all over the world

We may seize this opportunity to thank each and every one of the contributors to this successful Conference and Workshop for their invaluable contributions and for having travelled, sometimes from afar, to Lüneburg. Last but not least we want to thank Michael Valentine and Tatyana

Boldyreva for their superb translations and Florian Gliffe, Martin Gross and Timo Leder for helping to organize the conference

Prof. Dr. Vladimir M. Kolodkin
Prof. Dr. Wolfgang Ruck
Izhevsk and Lüneburg, in May 2004

1

INTRODUCTION

CHEMICAL WEAPONS CONVENTION AFTER THE FIRST REVIEW CONFERENCE

Jiri Matousek[1]
[1]*EU Research Centre of Excellence for Environmental Chemistry and Ecotoxicology, Faculty of Science, Masaryk University Brno, CZ-625 00 Brno, Czech Republic*

Abstract: The Chemical Weapons Convention is shortly characterised, and the main principles of, *inter alia* the General Purpose Criterion are stressed. The status of its implementation as for March 14, 2004 shows the main data obligatorily declared by 161 Party States and main achievements in destruction of Chemical Weapon (CW) stockpiles and destruction / conversion of CW production facilities and their verification. The Organisation for the Prohibition of the Chemical Weapons (OPCW) is briefly presented and the main results of the First Review Conference analysed.

Key words: CW Convention, Verification, CW destruction, OPCW, Review Conference

1. INTRODUCTION

The Convention on the Prohibition of the Development, Production, Stockpiling and Use of Chemical Weapons and on Their Destruction, shortly depicted as Convention on general and comprehensive prohibition of chemical weapons, or Chemical Weapons Convention, abbreviated as CWC, was adopted in 1992, after complex negotiations on the soil of Conference on Disarmament (and previous multilateral negotiating fora in Geneva), lasting nearly a quarter of a century, not only due to the existence, at that time, of East-West confrontation and the Cold War but mainly due to the worldwide spread of chemical industry and relatively easy possibility of clandestine synthesis of chemical warfare agents in militarily relevant quantities. This, as well as the bad experience with previously adopted Convention on the Prohibition of Development, Production and Stockpiling Bacteriological (Biological) and Toxin Weapons and of Their Destruction (opened for signature in 1972, entered into force in 1975), which lacked any objective verification mechanisms in the first line, has been reflected in very careful definitions and criteria, defining purposes not prohibited by the

V.M. Kolodkin and W. Ruck (eds.),
Ecological Risks Associated with the Destruction of Chemical Weapons, 7–14.

Convention and very complex and sophisticated verification system. The CWC is no doubt very impressive and the best elaborated disarmament document, totally outlawing one important and very dangerous kind of weapon of mass annihilation, committing the States Parties (SP) to the destruction of the chemical weapons (CW) stockpiles and production facilities (CWPF). Ten years after opening for signature and six years after entry into force (EIF), the First Conference of the SPs reviewing operations of the CWC has been convened in the The Hague, stating generally good acceptance by the international community, showing positive results of implementation of the CWC provisions, and defining the course for future.

2. CHEMICAL WEAPONS CONVENTION – BASIC FACTS

The Chemical Weapons Convention (CWC), opened for signature in Paris on January 13, 1993, entered into force on April 29, 1997. Its complexity is reflected in almost 200 pages of text, containing a Preamble and 24 Articles and three Annexes: On Chemicals (6 p), On Implementation & Verification (105 p), On Protection of Confidential Information (5 p) [1].

To the main pillars of the CWC:

- Verified destruction of chemical weapons (CW) and CW production facilities (CWPF), i.e. disarmament,
- Verified non-production of CW, i.e. non-proliferation,
- Assistance and Protection,
- International Cooperation.

The spirit of this Convention lies *inter alia* mainly in defining the scope of the prohibition. The CWC is rather purpose than compound oriented. This means that it is nothing like the list of prohibited compounds as some less informed people expect. The CWC′s leading principle, which is often reported as General Purpose Criterion (GPC) is contained in the wording of Article II, para 1, defining the purposes of the CWC as “Chemical Weapons”:

Article II: DEFINITIONS AND CRITERIA

For the purposes of this Convention:

1. “Chemical Weapons” means the following, together or separately:
a) Toxic chemicals and their precursors, except where intended for purposes not prohibited under this Convention, as long as the types and quantities are consistent with such purposes
b) Munitions and devices, specifically designed to cause…
c) Any equipment specifically designed for use…

Toxic chemicals are further defined in Article II para 2 as meaning

"any chemical which through its chemical action on life processes can cause death, temporary incapacitation or permanent harm to humans and animals. This includes all such chemicals, regardless of their origin or of their method of production and regardless of whether they are produced in facilities, in munitions or elsewhere".

From this explanation, quoting relevant articles of the CWC is evident, consistent with the mentioned GPC that the Convention:

1. is nothing like a list of prohibited compounds,
2. covers *any* toxic chemical intended to be used for chemical warfare (and therefore developed, produced and stockpiled), pursuant to Article II, para 1 (a) and para 2, this means even those not yet synthesised. This means that the CWC is open-ended and the prohibition covers any scientific and technological development.

The CWC lists (as the verification instrument) the most important toxic chemicals and their precursors, endangering the CWC within Three Schedules, constituted according to the risk the chemicals pose for the Convention. Schedule 1 contains super-toxic lethal chemicals and key precursors that have no peaceful uses, Schedule 2 contains less dangerous toxic chemicals and precursors produced in small quantities, and Schedule 3 lists toxic industrial chemicals (formerly used for chemical warfare) and precursors produced on a mass scale. A frequent misunderstanding occurs regarding the Schedules which list "prohibited compounds" although it is clearly stated in the CWC that "Schedules do not constitute a definition of CW". The open-ended prohibition, however, does not mean that new toxic chemicals (other than those contained in Schedules) cannot be used on battlefields by non Party States or less possibly by SPs breaching the CWC or more possibly by the terrorist groups. That is why the scientific and technological development is to be very vigilantly monitored, international verification measures extended, national authorities and operation systems established, and relevant legislation adopted in order to enable prevention and adequate response in real time (repression, protection, rescue and recovery) in cases of emergency.

3. STATUS OF IMPLEMENTATION OF THE CHEMICAL WEAPONS CONVENTION

(If not otherwise stated, the data on implementation are reported as of March 14, 2004).

- At present, there are altogether 161 Party States to the Convention. Important is the membership of all P-5 members of UN Security Council and the vast majority of states with declarable CWC facilities.
- Four SPs (Russia, USA, India and South Korea) declared possession of CW.
- Among SPs, there are 11 possessors of former (after 1946) CW production facilities (CWPFs), i.e. Russia, USA, India, South Korea, France, UK, China, Iran, Japan, Bosnia & Hercegovina and Serbia & Montenegro.
- The CWC implementation & verification regime now covers 90 % of the world's population, but more importantly, 98 % of the world's chemical industry.

Reviewing the figure on the number of SPs, it is also important to note that there are 21 signatory states that have not yet ratified (*inter alia* Israel) and altogether 12 countries that have not even signed. Beside not very important states it is necessary to note DPR of Korea and the neighbours of Israel (Egypt, Iraq, Lebanon and Syria) binding their signature on Israel's withdrawal from its nuclear weapons programme. One important breakthrough is the recent accession of Lybia.

Table 1: Universality: CWC compared with other main agreements on WMD

Treaty	Entry into force	SPs	other signatories	non-signatories
NPT	1970	187	---	7
BTWC	1975	151	16	27
CWC	1997	161	21	12

Assessing the universality of the CWC (by the way one of the requirement of the First Review Conference), one can come to interesting results comparing this requirement with the status of other principal agreements on weapons of mass destruction (WMD) as can be shown by table 1. It seems that one could be satisfied with a relatively high number of SPs, six years after EIF in comparison with other presented arms-control / disarmament agreements. Nevertheless, for the prevention of use of CW, it is necessary to reach a higher number of SPs, mainly because most of the above mentioned important non-SPs concentrated in the Near and Middle East and on the Korean peninsula are supposed to be nearly certainly possessors of CWs (not to mention possession of other WMD like in the case of Israel).

The worldwide status of CWC implementation is substantiated by other important data:

- 150 initial declarations (on possession / non-possession of CW) were obtained from SPs,
- 133 national authorities were established in the SPs.

Especially the latter number seems to be still insufficiently taking into consideration the tasks of such governmental office in the national implementation measures, starting with the respective legislation and then supervision of the domestic chemical industry and any cooperative activities with the Organisation for the Prohibition of Chemical Weapons (OPCW).

The most important data from the declarations of SPs (see table 2) show the worldwide problems associated with the possession, storage, former production of CW as well as with the spread of chemical industry as the point of outcome not only for the destruction of CW at present and in the near future, but for monitoring the non-production of CW in chemical industry in future.

Table 2: Important data from the declarations by the SPs (as of February 27, 2004)

Subject	Declaring SPs	Declared sites
CW storage facilities (CWSFs)	4	33
CW destruction facilities (CWDFs)	4	39
CW Production Facilities (CWPFs)	11	61[a]
Abandoned CW	3	15
Old CW	10	42
Schedule 1 Chemicals	21	27
Schedule 2 Chemicals	35	432
Schedule 3 Chemicals	34	509
Discrete Organic Chemicals	63	4490

[a] Of the 61 reported former CWPFs, 41 certified as already destroyed & converted

The total number of declared sites (5648) which are to be regularly or randomly inspected shows the high burden of verification activities to be expected. At this stage of implementation, the verification activities have been obviously concentrated on storage and destruction, and in industry on facilities producing scheduled chemicals. At present, the most important activity in implementation of the CWC is destruction of CW:

- declared chemical agents ~ 70 thousand m.t.
- destroyed (February 2004) ~ 8.4 thousand m.t.
- declared munitions (containers) ~ 8.6 M items
- destroyed (February 2004) ~ 1.9 M items

As expected, the destruction proceeds asymmetrically, meeting domestic financial and technological problems with construction of destruction facilities. So, e.g., the Russian Federation destroyed 400 tonnes (about 1 %) of its stockpiles as its promise to do so to the 1st Rev. Conf. It was expected that the scheduled 10 years' term for total CW destruction laid down by the CWC would not be ; the allowed exemption to extend the destruction period for another 5 years has been applied for. By the way, the USA did the same. Most important is however that the destruction process is going ahead.

4. ORGANISATION FOR THE PROHIBITION OF CHEMICAL WEAPONS – OPCW

Pursuant to the CWC, after its signature, the Preparatory Commission was founded and after EIF the Organisation for the Prohibition of Chemical Weapons (OPCW) with the seat in The Hague (Johan de Wittlaan 32, 2517 JR Den Haag) was established. For more information see http//:www.opcw.org.

The Organisation consists of three main elements:

1. Conference of the Party States (all SPs, meets regularly once a year), present
   Chair: H.E. Amb. Dato′ Noor Farida Ariffin (Malayisia).
2. Executive Council (41 members distributed among the SPs on a regional, rotating base for a 2 years' term, meets regularly 4 times a year)
   present Chair: H.E. Amb. Petr Kubernát (Czech Republic)
3. Technical Secretariat (492 staff members, of them about 200 inspectors), Director General: H.E. Amb. Rogelio Pfirter (Argentina),
   Subsidiary bodies: Scientific Advisory Board (20 independent experts), Confidentiality Commission, Advisory Board on Administrative and Financial Matters.

5. THE FIRST REVIEW CONFERENCE

The character and tasks for the Conference were determined as follows:

- Review operations of the Convention,
- Take account of scientific and technological development,
- Lessons learned and recommendation for future implementation,
- Not an amendment (revision) conference.

The attendance represented 113 SPs, 2 signatory states (Haiti, Israel), 2 non-signatory states (Libya, Angola), 5 International Organisations (ESA, ICRC, PCA, CTBTO, UNIDIR), 22 NGOs and 6 Industry Associations. Despite provocative statement by the US alleging non-compliance by Iran and concerns about the Sudan, the Conference did not collapse into disarray; the CWC has not met the fate of BTWC. The Conference did not result in radical change of direction for the OPCW or substantive decisions on crucial, still outstanding issues (e.g. so called "non-lethal" agents, riot control agents, "law enforcement", nil declarations in respect of OCPFs and the like. A number of priorities have, however been clearly recognised. To those priorities belong:

- Universality of the Convention,
- National implementation measures,
- International Cooperation and Assistance,
- Verification regime for the chemical industry
- Optimisation of verification measures
- Scientific and technological development and
- Functioning of the OPCW.

The detailed explanation goes beyond the frame of this paper. For further information see the documents adopted. This is mainly the Political declaration containing 23 paras [2] and the main written result, i.e. the Review document with 134 paras [3]. Except many statements, mostly only general, the programme did not go too deeply into the problems of the impact of scientific & technological development on the CWC that are obviously connected with its future implementation. This problem was analysed in the document prepared by the OPCW Scientific Advisory Board introduced in the Note by the Director General [4]. It is generally expected that this will mainly influence the future activities of the OPCW.

6. CONCLUSIONS

The operations of the Chemical Weapons Convention are proceeding satisfactory, judging by the status of its implementation by 161 Party States and verification by the Organisation for Prohibition of the Chemical Weapons in The Hague seven years after entry into force. The First Review Conference (May 2003) stressed the importance of achieving worldwide universality in order to totally liquidate the legacy of past chemical arsenals once and for all, prevent threats and utilise benefits of the scientific and technological development for the CWC implementation in the foreseeable future.

References

[1] UN (1993). Convention on the Prohibition of the Development, Production, Stockpiling and Use of Chemical Weapons and on Their Destruction.UN, New York.
[2] OPCW (2003). Political declaration http://www.opcw.org/cwcrevcon/doc/NAT/FRCPoliticalDeclaration.html OPCW, The Hague.
[3] OPCW (2003). Review document. www.opcw.org/html/global/wgrc/2k3/rc1revdoc.html. OPCW, The Hague.
[4] OPCW DG (2003). Note by the Director General: Report of the SAB on Developments in Science and Technology, OPCW, Conference of the SPs, RC-1/DG.2, 23.04.2003. OPCW, The Hague.

IMPLEMENTATION OF RUSSIA'S OBLIGATIONS TO DESTROY CHEMICAL WEAPONS IN THE UDMURT REPUBLIC

Anatoly A. Fominykh[1]
[1]*Chairman of the Committee on Conventional Problems of Chemical Weapons under the Government of the Udmurt Republic, Izhevsk (Russia)*

Abstract: The territory of the Udmurt Republic (UR) there are two arsenals with chemical weapon (CW). The total volume of stored chemical warfare agents is more than 12,000 tons (more than 30% of the Russian CW stockpile) which the Russian Federation (RF) is to destroy in compliance with the international obligations of the Convention on the Prohibition of CW. The safety not only of the Udmurt Republic but also the contiguous regions - Tatarstan, Bashkortostan and the Kirov Oblast - depends on the safe storage of the chemical warfare agents and timely and competent destruction of these arsenals. It has been almost six years since Russia joined the Convention on Prohibition of CW. During this period Russia has been able to advance along the path of chemical disarmament. The advance is perhaps not as far as we wish, but the positive dynamics are evident. Yet, many issues connected with the provision of environmental safety during work with CW have not been solved, or are in the first stage of their realization. The objective of the Government is to prevent emergency situations of global character while carrying out this work, the consequences of which – according to the experts' appraisal – may be significant losses of population and extensive damage to the environment.

Key words: convention on prohibition of chemical weapon, chemical weapons, environment, ecological safety, chemical weapon destruction

The basic objective of the democratic state is to create the most favorable conditions for its citizens' rights and enjoyment of freedom.

The Constitution of the RF (Article 72) specifies a range of objects which are under the jurisdiction of the state and the subjects of the RF. In particular, they include protection of the rights and freedom of man and citizen, the environmental protection and provision of the ecological safety, and implementation of the international treaties of the Russian Federations.

One of the international treaties signed by the RF in January 1993 is the Convention on the Prohibition of the Development, Production, Stockpiling and Use of Chemical Weapons and Their Destruction [1]. Russia was one of the first to sign this international treaty, thereby demonstrating to the

V.M. Kolodkin and W. Ruck (eds.),
Ecological Risks Associated with the Destruction of Chemical Weapons, 15–23.

international community her intentions to continue successively carrying out the policy of turning away from all kinds of weapons of mass destruction.

Chemical weapons have always been covered with a thick veil of mystery and plunged in an informational vacuum. It is understandable: in the period of great opposition - "cold war" - chemical weapons were one of the important elements of national security of the great power called the USSR, as well as of others. Chemical weapons were taken into service by the armies of the most developed countries of Western Europe, China, Japan and, of course, the USA.

The military opposition of two political systems resulted in stockpiling huge amounts of CW. At present, the stock of CW in the arsenals of not less than 15 countries, according to experts' appraisal, is over 100,000 tons, being second by its volume only to nuclear weapons [2].

Officially production of CW in the USSR stopped in 1987, in the USA in 1969 [3].

It should be noticed that the decision Russia made to stop production of CW is undoubtedly in agreement with Article 7 of the RF Constitution which states that Russia is a social democratic state whose policy is to create the conditions ensuring worthy life and free development of citizens. In essence this definition may be considered as the goal of the state. In some researchers' opinion one of the most important and primary goals of the state is to prevent its population from dying out. It does not matter what reasons and circumstances may lead to the death of a nation (or part of the nation), the goal of the state is to foresee it and take timely measures to prevent collapse. In this case the goal is to prevent emergency situations of global character while storing and destroying the stockpile of CW which, according to experts' appraisal, may result in significant losses of population and extensive damage to the environment.

The fundamental document in the area of CW destruction is the above-mentioned Convention on the Prohibition of the Development, Production, Stockpiling and Use of Chemical Weapons and Their Destruction ratified by Russia in November of 1997.

It has been almost six years since Russia joined the Convention on the CW prohibition. During this period Russia has been able to advance along the path of chemical disarmament. The advance is perhaps not as far as we would like, but the positive dynamics are evident. The first years of realization of the chemical disarmament program coincided with political and economical reforms in Russia, which undoubtedly affected the course of implementation of the program measures and led to the lag of four years from the scheduled terms, minimum [4].

According to the opinion of the former state buyer of the program (RF Ministry of Defense), the main cause of the program's lagging behind lies with the lack of financing for the total complex of work in the field of the CW disarmament [5]. The aforementioned cause was confirmed by the conclusions of special inspections done by the RF Security Council and Central Control Board under the RF President in 1998-2001 on the orders of the RF President.

For the last two years the President and the Government of Russia have taken a number of fundamental measures as a result of which the negative tendency was overcome.

By now RF has a normative legal basis in the area of CW disarmament formed in general; basic federal laws regulating the relationship between work with chemical weapons and the social protection of the population during this work have been passed. In accordance with the basic federal law in the field of CW disarmament "On Ratification of the Convention on the Prohibition of the Development, Production, Stockpiling and Use of Chemical Weapons and Their Destruction" [6] a unified system of state management of the CW disarmament process and implementation of the Convention has been created. The RF President's decrees have defined a new state buyer of the Federal Special Program "Destruction of the Chemical Weapon Stockpile in RF" which is the Russian Munitions Agency [7]; and a State Commission for Chemical Disarmament was established [8]. In 2001 a new program of CW disarmament in Russia was worked out and approved in addition to the program accepted by the RF Government in 1996 [9]. In essence a new concept of the program of CW disarmament in RF was developed.

It should be noted that the new concept of the program differs fundamentally from the previous one – financial expenses and the terms of liquidation (or conversion) of the former producers of CW have been considered; the number of established full-scale facilities for CW destruction has been reduced; the terms of construction and operation of the facilities for CW destruction have been corrected; transportation of munitions with chemical agents from the arsenal in the Kizner area (Udmurtia) to the destruction facility in the Kurgan Oblast has been planned; for the personnel involved in the principal operations on CW destruction the shift mode of work has been proposed; the volumes of financing of capital investments to construction of the social sphere facilities have been increased up to 10 % from the cost of the construction of the facilities for CW destruction, and a range of other suggestions has been made. The estimated cost of the program is 94.3 billion roubles (2001 values).

The strategy of the new program is built taking into account the position of the regions, public opinion, opinion of the international community, and, most importantly, the main partner in CW disarmament, the USA, which signed a bilateral treaty on cooperation with Russia in support of the Russian program as far back as 1992. Some amendments were introduced into the basic Federal Law "On Chemical Weapon Destruction" [10] which allowed transportation of chemical weapons within Russia's boundaries.

RF has started making real steps on CW destruction. First of all, the chemical weapons of the 1st and 2nd categories were destroyed in 2001. Second, in December of 2002 the first Russian facility for CW destruction was constructed and was put into operation in Gorny, Saratov Oblast. On April 26, 2003 Russia reported to the world community on the implementation of the first stage of her international obligations according to the Convention on the Prohibition of Chemical Weapons – the completion of destruction of 1% out of the total of the CW stockpile declared by Russia at the time of signing the international treaty. We also significantly advanced both in disposal and conversion of the former facilities of CW.

Summarizing the above, it should be noted that after the presidential elections new Russian President Putin thoroughly acquainted himself with the course of establishment of the CW disarmament program, which has the status of the Presidential program.

As a result of getting acquainted with the causes of non-implementation of the international obligations of Russia on the CW disarmament, President gave relevant instructions to the RF Government and the Council of Defense of RF. Issued were decrees on the improvement of the management system of the CW disarmament; the funding for the expenses for the accomplishment of measures of this program by the Federal Treasury was significantly increased, namely from 500 million Roubles in 1988-2000 to 3 billion Roubles in 2001 (6-fold!). In 2002 the volume of the budget meant to finance the program was 5.3 billion Roubles. For 2003 the Federal budget had the same assignment of 5.3 billion Roubles for the implementation of the program measures.

Despite the experience of the USA, the process of creation of the Russian system of state management of the CW disarmament process was difficult, which was connected with rather serious political and economical problems of the new democratic state formation.

The arsenals of the CW stockpile are in six objects of RF – Bryansk, Kirov, Kurgan, Penza, Saratov Oblasts and the Udmurt Republic (UR). The major stockpile of CW is in the UR – more than 12 tons representing 30% of the Russian stockpile (Kambarka and Kizner). Another 6960 tons of CW (17.4%) is in the neighboring Kirov Oblast (the Maradykovsky settlement).

The situation puts the region in a very difficult situation and demands special measures within regional policy regarding the republic participation in the CW destruction and, in particular, as to ensuring the ecological safety while storing and destroying CW, preventing possible emergency situations at the arsenals of CW storage.

The facilities of CW destruction are supposed to be located in the vicinity of small towns and villages, which are characterized by the lack of industrial development and social infrastructures, unbalanced local budgets, and low standard of living. In addition, all areas are subsidized. Due to this fact expenditures will be needed for the development of power-systems, construction and reconstruction of transportation systems, monitoring the environment and public health, establishment of environmental protection measures, and outreach with the population to form favorable public opinion on the problems of CW destruction.

To realize the Federal Special Program (FCP) "Destruction of Chemical Weapon Stockpile in RF" on the territory of the UR a resolution was made by the Administration of the UR on April 30, 1997 #467 which confirmed the division of the responsibilities among the Republic executive authorities during the conduction of the work on the construction of the facilities for CW destruction [11].

Twelve governmental bodies of the republic are involved in the implementation of this resolution, the main ones being the Ministry of Natural Resources and Environmental Protection, the Ministry of Health Protection, the Ministry of Construction, Architecture and Housing Policy, Ministry of Internal Affairs, the Ministry of Civil Defense and Emergencies, and the Republic Center of State Sanitary and Epidemiological Inspection.

The State Council of the UR approved a resolution on October 13, 1998 #681-1 "On Particularities of the Application of the Federal Law in the Udmurt Republic "On the Chemical Weapons' Destruction" which prohibits the importation of CW and wastes formed as a result of the destruction process, used equipment from the territories of other subjects of RF and foreign countries. The resolution determines the mechanism of coordination of location of CW destruction facilities sites, grounds for the burial of wastes and routs of transportation of CW to the places of destruction.

In the period of 1992-2002 the government bodies of the UR made over 40 resolutions and orders concerning different aspects of storage and destruction of CW on the territory of the UR.

The need for elaborating and adopting these normative acts was dictated by the necessity to develop and supplement Federal standard legal acts considering local peculiarities.

Taking into account the complexity and ambiguity of the problem, the government bodies of the UR solve practically all issues that are under the jurisdiction of the RF subject together with the local government bodies [12].

During the period of preparatory work for the CW destruction process on the territory of the UR over 500 million Roubles were received from the Federal budget to realize the measures of the Program.

The grant was spent on the design and constructinon of the facilities of the social sphere in the Kambarka and Kizner areas. In Kambarka area, the construction of the gas pipeline to Kambarka is being completed; reconstruction of the gas fuel boiler-house will be completed soon in the northern housing area of Kambarka; a 60-unit apartment house is finished and two similar houses are being built; a hostel for 96 people is finished; the Central Republican Hospital has been reconstructed; a canal through Kambarka lake was built; the Kambarka-Mikhailovka highway is under construction; the construction of a medico-diagnostic center was started as well as a bus station for 50 passengers; the beginning of construction of the town canalization is planned. In the Kizner area a district hospital with clinic was built and placed in operation in the Bemyzh village; in the district center, water and gas pipelines are being built; in 2003 the construction of a 500-seats concert hall began.

And finally, in the middle of this year the state buyer of the program approved the technico-economic ground for the construction of the industrial zone of the facility for CW destruction in Kambarka. The builders started their first-priority works and reconstruction of the lewsite storage sites. Putting into operation the facility for CW destruction in Kambarka is planned for the end of 2005.

Up to present the unified system of monitoring of the environment and health of the population who live in the area of protective actions has not been developed though local and federal scientists and specialists have carried out selective investigations of the environment in the Kambarka area and health condition of the population in the Kambarka and Kizner areas. The indicators of the state of public health are similar to those of other areas of the republic. The ecological situation of these areas does not differ from the state of the environment in the republic as a whole.

Taking into account the condition of the Russian economy, the implementation of the requirements of the Convention does not seem possible without international assistance.

In the UR, thanks to the financial support received from Sweden, the work on the prediction of after-effects of emergency situations possible during storage of CW in Kambarka was carried out; computer equipment was supplied with software on the prediction of emergency situations and a system for operational decision-making to warn the personnel and population of the area; medical equipment for the Central Hospital in Kambarka has been supplied; a Public Relation Center has been formed and established in Kambarka. Finland supplied the equipment for environmental monitoring on the territory of the base of CW storage in Kambarka. Specialists of the Center of State Sanitary and Epidemiological Inspection of the UR and the Center of Environment and Occupational Diseases of the

National Hospital of Oslo with the financial support of a non-governmental organization of Norway in 2000 carried out the first stage of the research on the evaluation of the environmental condition in Kizner area [13] in connection with the long-term storage of chemical weapons.

It is not clear yet how far the partners of Russia are ready to go to develop cooperation in the area of chemical disarmament. At the same time, the important representatives of western governments understand that to pursue the interests of their own safety and international stability it is necessary to help Russia to dispose of the arsenals of CW and to participate in this process financially. One of the objectives of the local authorities and the bodies of municipal management is to define specific projects and tendencies for the purposeful and efficient application of the financial and material foreign assistance and to coordinate these suggestions with the plans of the federal bodies responsible for the work with the states–donors.

References:

[1] Convention on the Prohibition of the Development, Production, Stockpiling and Use of Chemical Weapons and Their Destruction. (1994) Moscow Journal of International Law. 3:170-207

[2] Lisov, O.M. (1994) International Law Documents on Limitation, Prohibition and Liquidation of Chemical Weapons – Weapons of Mass Destruction. M., pp 72

[3] Sheluchenko, V.V., Demidiuk, V.V. (2000) World Community and Chemical Weapons. M., pp. 9

[4] Kharitchev, A.D. (2003) On the Work of the State Commission on Chemical Disarmament. Public Forum-Dialog "Russia's Implementation of the Convention on the Prohibition of Chemical Weapon: Condition and Prospects by the end of 2002." (2003) 'Agenstvo Rakurs Production' Ltd. pp 9

[5] Petrov, S.V. Chemical Weapon Destruction: National Problem and Conventional Obligations. (1999) Federal and Regional Problems of Chemical Weapons Destruction. Informational Collection. Russian Institute of Scientific and Technical Information. M., pp. 7

[6] On the Ratification of the Convention on the Prohibition of the Development, Production, Stockpiling and Use of Chemical Weapons and Their Destruction: Federal Law as of November 5, 1997 #138-FZ. (1997) Code of RF Laws. #45, art. 5138

[7] On the Structure of the Federal Executives Bodies: RF President Decree as of May 25, 1999 # 651. (1999) Code of RF Laws. #22, art. 2727

[8] On the State Commission on Chemical Disarmament: RF President Decree as of April 26, 2001. #487. (2001) Code of RF Laws. #18, art. 1828

[9] On Changes and Amendments to the RF Government Resolution of March 21, 1996 #305 "On Ratification of Federal Purposeful Program 'Destruction of Chemical Weapons Stockpile in the RF": RF Government Resolution as of July 5, 2001 # 510. (2001) Code of RF Laws. #29, art.3020

[10] On Chemical Weapon Destruction: Federal Law as of May 2, 1997 #76-FZ. (1997)) Code of RF Laws. #18, art.2105

[11] On UR Participation in the Federal Purposeful Program Realization "Destruction of Chemical Weapons Stockpiled in RF": UR Government Resolution of April 30, 1997 #467

[12] On Changes and Amendments to the UR Government Resolution of April 30, 1997 #467 "On UR Participation in the Federal Purposeful Program Realization "Destruction of Chemical Weapon Stockpile in RF": UR Government Resolution # 704 of August 19, 2002

[13] Chemical Safety. Environment and Public Health. (2001) First International Conference. Theses. "Udmurtsky Universitet." Izhevsk. pp156

THE CWC AFTER THE REVIEW CONFERENCE

Walter Krutzsch[1]
[1]*Deputy Head of the Delegation of the GDR to the CWC Negotiations*

Abstract: The Convention on the Prohibition of Chemical Weapons (CWC) supplements the Geneva Protocol of 1925. In order to exclude any use of chemical weapons it prohibits any possession of chemical weapons and installs a strict verification system. In the Review Conference States Parties pledged to increase their efforts to this end. When this momentum shall be kept, existing shortcomings are to be overcome. This relates to the following subjects: infringements into independence and integrity of the Technical Secretariat (TS); living up to the undertakings of the States Parties as defined in the Conventions text; adaptation of the verification system to changes in science and technology; scrapping the recently adopted tenure policy of 7 years maximum for staff and inspectors and, instead, ensure that the TS will retain its experienced people; public accountability of the Organisation for the Prohibition of Chemical Weapons (OPCW) concerning results in the endeavour to exclude completely the use of chemical weapons; strengthening the effectiveness and transparency of work of the political organs of the OPCW, especially the Executive Council; enhancing co-operation between States Parties especially in the destruction of CW stocks.

1. INTRODUCTION

The concept of the CWC evolved during the cold war period. The threat of mutual annihilation in a nuclear holocaust was widely considered a necessary element of stability. The Biological Weapons Convention (BWC) and the Chemical Weapons' Convention (CWC) were part of attempts to contain fatal consequence of the nuclear arms race. The use in war of both, biological and chemical weapons, was, and still is, prohibited by the Geneva Protocol of 1925. But prohibition of use alone had proved to be unreliable if not supplemented by a norm ruling out any form of possession of those weapons and preventing their reappearance. In the sixties and seventies of the last century, the danger emanating from chemical weapons was considered imminent in contrast to biological weapons. Therefore, western participants in the negotiations considered effective verification to be the condition for a reliable verified prohibition of CW.

The logic for such a position was: Security interests may require upholding a CW stock and the right to retaliate a CW attack by a State Party in violation of the Convention or by a State not Party. Therefore, to renounce

V.M. Kolodkin and W. Ruck (eds.),
Ecological Risks Associated with the Destruction of Chemical Weapons, 25–38.

unconditionally such option presupposes reliable guaranties that clandestine existence and reappearance of CW is world-wide excluded. Finally, the inclusion of an effective verification system into the CWC became a generally recognised goal. In 1992, negotiations entered into their decisive phase and were crowned with success. The CWC, an instrument of international law, will replace a weapons' system.

2. TASKS TO BE RESOLVED

(The following is intended to supplement and support the paper of Professor Jiri Matousek: "Chemical Weapons Convention after the First Review Conference")

Not all is bad with the implementation of the Convention. We have 161 States Parties and have conducted verification activities under Articles IV, V and VI in 62 States Parties with about 1600 inspections or inspection equivalents. But that there are things we should concern ourselves with in order to ensure that the letter and spirit of the Convention are met. In the light of the Political Declaration of the Review Conference [1] further efforts should be directed to achieving the object and purpose of the Convention.

2.1 Independence of verification activities

The States Parties stress in paragraph 12 of the Political Declaration that the verification system shall be applied in a non-discriminatory way and (in paragraph 13) the importance of and their commitment to an effective and credible verification regime.

The preconditions for reliability and credibility of verification results are multilateralism, equality of rights and obligations of the CWC States Parties and independence of the Director-General and the staff members of this Organisation. Article VIII, paragraphs 46 and 47 of the CWC must be scrupulously complied with. Those provisions prohibit the Director-General, the inspectors and the members of the staff "to seek or receive instructions from any Government or from any other source external to the Organisation". At the same time the States Parties are obligated to "respect the exclusively international character of the responsibilities" of those persons and "not seek to influence them in the discharge of their responsibilities". If those provisions were ignored and the principle of independence of the Director-General, the inspectors and the other members of the staff of this organisation undermined, this would open up the verification system to be damaged by power policy and economic rivalries.

Without those safeguards, verification will be at the pleasure of "key delegations", especially those of the main contributors to the Organisation's budget. That would mean: those countries would only be verified at their convenience and nothing would be found what they do not wish to be found. Their predominant influence on the Secretariat's activities would undermine multilateralism, equality of rights and obligations, and credibility of verification results.

The Political Declaration could be an opportunity to discontinue practises criticised by the Administrative Tribunal of the International Labour Organisation [2]:

"In accordance with the established case law of all international administrative tribunals, the Tribunal reaffirms that the independence of international civil servants is an essential guarantee, not only for the civil servants themselves, but also for the proper functioning of international organisations." The Tribunal warned against "violation of the principles on which international organisations' activities are founded (and which are in fact recalled in Article VIII of the Convention, in paragraphs 46 and 47)".

In order to preserve independence of the TS in its function as the verification organ of the OPCW the provisions on privileges and immunities of the members of the inspection teams are set out in Part II, 'General Rules of Verification', Section B, of the Verification Annex. The privileges and immunities stipulated in Section B of this Part provide full diplomatic immunity. The Convention uses the 1961 Vienna Convention on Diplomatic Relations as the basis for specifying privileges and immunities. This sets a new standard for international organisations which have on-site inspection activities to conduct on the territory of member States [3]. These key provisions for effective verification have been weakened in many instances, e.g.: The inspectors' notebooks are protected by immunity as stipulated in Part II, Section B of the Verification Annex of the Convention. However, the implementation of this provision is withheld by misinterpretation.[4] This misinterpretation is rooted in the concept of a decision of the Conference of the States Parties adopted 1997 (CI/DEC.51).[5]

The Technical Secretariat has been micro-managed by States Parties since the Convention entered into force. For example, the verification efforts are tightly controlled by the budgetary process. The perceptions of risk to the Convention as derived by the Technical Secretariat from declarations, inspection information and experience are not the determining factors in planning the verification process. Therefore, Schedule 1 facilities and the Schedule 2 plant sites based often are multiply controlled in over 40% of the inspections in 2003 while the TS has inspected only 25% of Schedule 3 plant sites and only a little over 1% of "Other Chemical Production Facilities" according to Part IX of the Verification Annex of the Convention. For the year 2002 the estimated cash income available for the OPCW stood at EUR 58 million as compared to the approved budget of EUR 61,9 million. Full programme delivery in 2002 would have required a budget of EUR 64,1 million. The deficit for 2002 was caused by an under-budgeting by the member-states for at least three consecutive years. In this time period a policy of zero growth was imposed. The decision of member states had ignored the increase in staff-related costs and the fact that new verification tasks required additional costs. In this period of time, the first inspections of the US chemical industry became due. In addition, Article IV and V of the Convention verification costs for CW and CWPF destruction incurred for the Organisation were not immediately reimbursed by the inspected State Parties. This raised additional cash problems for the Organisation. Therefore, due to financial pressures the Technical Secretariat accomplished 2001only 67 per cent of the approved number of inspections and calculated that it could only conduct just over 50 per cent of the inspections originally approved for 2002. It is highly disturbing that, in a period of increasing danger by the threat of weapons of mass destruction, the political organs of the OPCW did not solve the financial problem of the Organisation for a third year. Key member states like USA, Japan and Germany were, because of their strength and influence, more than others responsible for this lamentable budgetary situation.

The problem was overcome not without help from outside. A signal came from Brussels. The Council of the EU agreed April 2002 that the EU and its member States will:

- promote compliance with obligations and commitments including the destruction of prohibited weapons, the prevention of their dissemination and illegal use, as well as the prevention of dissemination of their technologies;
- enact and strictly apply national implementation legislation;
- review the financial resources required by the international organisations in order to provide sufficient funding to enable them to discharge their monitoring activities, including those undertaken in the light of the new threats post September 11 and ensuring that the funds provided are used in the most effective way;
- sustain and expand the OPCW capabilities to conduct effective inspections, especially challenge inspections and investigations into alleged use. More realistic and frequent training exercises especially practice inspections, provide an ideal mechanism to maintain and enhance such capabilities.

The ability of the TS to use sampling and analysis equipment on inspections has been severely curtailed. Some States Parties prefer to have their own equipment used and hinder the use of approved equipment of the

TS. According to the General Rules of Verification, paragraph 27, there shall be no restriction by the inspected State Party on the inspection team to use orderly approved equipment which the TS considered necessary for an inspection. The analysis of samples was considered to be a very important tool provided by the Convention to obtain unquestionable evidence of compliance and non-compliance. The implementation of this tool was impeded by several means. This started in 1997 with the legally flawed reservation of the USA not to allow samples analysed outside their territory. India followed this bad example and included this denial into national law. The obstruction of the relevant Convention's provision continued with financial constraints for equipment and training of personnel imposed by the Executive Council. Nowadays, some States Parties are questioning the need and even the right of inspection teams to take samples. The reaffirmation of the undertakings in the Convention gives an opportunity to remedy this substantial flaw. If the present situation prevails, it could end in sampling and analysis capability of the TS being eliminated under the pretext that it is ineffective to maintain the capability for something that is not used.

2.2 Developments in science and technology

The Review Conference recognised the need to consider new industrial methods, such as micro and nano-reactor technologies not considered by the Convention with the potential of these reactors. (Large amounts of toxic agents can be made in reactors that would fit into a briefcase.) This will require great attention to find an early solution for appropriate verification measures. It is important to adapt the verification system quickly to changes in science and technology happening in industry and elsewhere. Examples: new chemical technologies and the need to look more closely at certain facilities producing unscheduled discrete organic chemicals (DOC); inclusion of data on non-scheduled risk chemicals into OPCW analytical database. Agents of the Novichok-category are not in the public domain and also not controlled through the Schedules. The OPCW should be aware of them and be capable of finding them in samples.

Other new dangerous developments in science and technology also require political attention. One year ago, information appeared about US research and development programs on non lethal chemical weapons that violate the CWC. The 'Sunshine-Project' reported on such activities on anaesthetics and psychoactive substances, development of long-range military delivery devices for these chemicals, including an 81mm chemical mortar round. Their intended use is against "potentially hostile civilians", in anti-terrorism operations, counterinsurgency, and other military operations [6]. The hostage drama in Moscow in which fentanyl, an anaesthetic, was used, thereby killing more than 120 hostages, was considered one of the scenarios for which such new weapons are needed. Speculations about an envisaged use in war of those weapons stirred up discussions that put in doubt the legal dividing line between the use of chemical weapons on one hand and law enforcement including riot control purposes on the other. The CBW Conventions Bulletin dedicated the editorial of its September 2003 issue to so-called non-lethal weapons. The highly important article considers recognition of the threat they pose as increasingly urgent since, "investment mounts in emergent 'non-lethal weapons' technologies" [7]. At the same time propaganda is busy winning more support for the false assumption that those weapons became unconditionally necessary. In response to this, the Bulletin points to the danger emerging from this threat. The euphemistic title 'non-lethal' is meant to deceive about their toxicity. What is more, it is meant to divert from the historic experience that any use of lethal chemical weapons was prepared with the use of 'non-lethal' ones. With other words: Their appearance in military arsenals and battlefields could have lethal consequences for the Chemical Weapons Convention. In preparation for this, pseudo-legal argumentation is used pretending those agents are not covered by the prohibitions of the CWC. The opposite is true. Their toxic properties are transgressing the strict borderline exempting some chemicals, e.g. tear gas for law enforcement and riot control purposes from prohibition. A study presented to a workshop in spring 2003 [8] concludes:

Any toxic agent intended to be used for law enforcement including domestic riot control must have properties consistent with those defined in paragraph 7 of Article II of the CWC. If this is not the case, it is prohibited and has to be declared and destroyed as all chemical weapons. The paper concludes with:

"9. Further action

Consideration and action by the States Parties, the policy-making organs of the OPCW, especially at the CWC Review Conference, as well as public discussion on this subject are required:

The Review Conference should

- reaffirm that the prohibitions of the CWC cover all chemicals regardless of their origin or method of production;
- recall that die CWC complements the obligations not to use such weapons, assumed under the Geneva Protocol of 1925 and
- warn against any violation of these norms."

Nowadays, it is hard to learn from The CBW Conventions Bulletin's editorial: "The First Review Conference earlier this year was opportunity to address the issue constructively. But, save in the national statements of New

Zealand, Norway and Switzerland, the OPCW chose not to do so. In the programme of Review-Conference follow-up work that is now getting under way, there is no mention of disabling chemicals nor even tear gas, still less the so-called calmatives and other such incapacitating agents in which interest is now rapidly reawakening."

2.3 Effectiveness and efficiency of the Verification system

The meetings of the Review Conference had been suspended for a day and the delegations of the States Parties convened in a Special Session and adopted a decision on a TS tenure policy of 7 years maximum for staff and inspectors. The first period is to begin 1999. The decision insists on a turnover every year of 14.5% of all personnel belonging to these categories. This will ensure that the TS will not retain its most experienced people, particularly its technical people in Verification and the Inspectorate. The corporate memory of the Technical Secretariat built up since 1993 will be gone by 2009. The increasing load to inspect CW destruction facilities coupled with the need to train new people will not only greatly degrade the level of professionalism and integrity of the entire verification effort, but also their frequency and intensity required (Declaration paragraph 14). The Executive Council had dealt with the tenure policy issue since years, since there was awareness of the dangers related to such concept [9]. There is striking inconsistency between such a decision and, the other day, a commitment to a credible and effective verification regime related to CWs and their destruction (paragraph 13 of the Declaration) and related to the chemical industry (paragraph 14). Having barely overcome the damage self-inflicted by the financial crisis on the verification regime, the next, even heavier, damage has been brought on its way with the tenure policy decision.

An increase of efficiency and effectiveness is foremost needed in the work of the Executive Council. The resolution of unresolved issues including some left over from the Paris Resolution of 1992 are discussed and discussed with very few decisions being made. The record of the Executive Council is less than encouraging, especially, when it comes to the solution of key issues of the verification regime. The number of unresolved issues has not diminished. This goes for both, destruction of chemical weapons and industry verification. The stock of unresolved verification issues currently under discussion, most of which since 1997, includes: – The meaning of "primarily for the development of CW" (in Article III (c) (vi));

- Guidelines to determine the usability of CW produced between 1925 and 1946;[10]
- Destruction/verification requirements for old and abandoned chemical weapons;
- Attribution of the costs related to inspections of old chemical weapons;[11]
- Extension of Russia's destruction deadlines;
- Assistance and protection against chemical weapons;
- Verification at Schedule 1 facilities;
- Restrictions on the transfer of Schedule 3 chemicals to States not party to the Convention;
- Sampling procedures;
- Low concentration limits for declaration of mixtures of chemicals containing Schedule 2A and 2A* chemicals;
- the endpoint of destruction;
- low concentration limits for Schedule 2A and 2A* chemicals;
- implementation of Section B of Part IX, especially paragraph 11(c);
- transfer of Schedule 3 chemicals to States not party to the Convention;
- how to declare captive use, etc.

Furthermore, there has been a significant delay within the Executive Council in the approval of detailed plans for the verification of destruction of chemical weapons or chemical weapons production facilities and/or decisions on requests for the conversion and plans for the conversion of chemical weapons production facilities. This has caused serious difficulties and led to increased costs in OPCW verification activities. At least two States Parties (Russia and the United States) had to begin destruction or conversion operations in the absence of approved detailed plans, with continuous monitoring by OPCW inspectors, which generates higher inspection costs [12]. A bright spot was the recent decision on boundaries of production, which greatly enhances the potential for even-handed declaration of the production of Schedule 2 and Schedule 3 chemicals.

Decision making *only* by consensus as a rule is inconsistent with the Convention and has proved to be counterproductive. Decision making as laid down in the Convention would inspire creative contributions and overcome stalemate by mutually agonising linkages. At present, results, if any, will confine the EC to register the lowest standard.

2.4 Cooperation between State Parties in destruction of CW stocks

The security interests of all State Parties require the destruction of all chemical weapons under strict international control in accordance with the necessary environmental security requirements. The Convention obligates the Possessor State to carry out the destruction and pays for the verification costs incurred. The co-operation of States Parties is a general principle of the Convention that influences all parts of activities necessary for achieving the object and purpose of the Convention.

The destruction of chemical weapons in general turned out to be more burdensome and expensive than expected. Russia, the USA and South Korea have asked to extend destruction deadlines. Financing destruction became an especially difficult problem for Russia. This has been recognised by the OPCW. It was decided to engage the Organisation to co-ordinate international co-operation including financial and technical assistance for this purpose. The estimated costs for the Russian destruction programme are huge. The USA became the main foreign sponsor. Germany is second to USA in sponsoring, followed by other EU countries. According to press reports, Switzerland and other countries recently announced further contributions. States Parties have indicated their readiness to assist Russia in destruction of its CW stocks. The slow progress in the implementation of the Russian programme should motivate the increase of such assistance. Co-operation and assistance might become more effective when rendered within a common programme. Possible advantages of such a programme might be better co-ordination of sponsors. Another positive impulse could be that the help for the destruction of Russian CW becomes understood by the politics and public for what it really is: An undeniably necessary and life-saving task.

2.5 Transparency of activities and accountability for results

The Political Declaration of the Review Conference states in its first paragraph that the Convention, universally and effectively implemented, will be an asset for all humankind. For this statement to be more than a good phrase, it must be followed by a concept of action for public information. The OPCW is not transparent. Except for a sanitised Annual Report, not much reaches the public, including NGOs. The annual Verification Implementation Report is classified 'highly protected' and remains within the TS and the States Parties governments under lock and key. The last unclassified report was issued several years ago, when the Director-General announced that States Parties not responding to requests for solving uncertainties about compliance will be named in the report. The Executive Council's or Conference's documents are normally not available, forget about public attendance to the session of these organs. The OPCW web-site only published press releases, vacancy notices, the Convention's text and general descriptions of OPCW activities. There is no way for anyone else to find out how things are going. The activities of the Organisation and their success and unresolved problems are widely unknown.

In his first statement the new Director-General of the OPCW promised transparent action vis-à-vis the States Parties. This is a laudable intention. But under the present circumstances, substantive information released from the OPCW Headquarters reaches only the delegate(s) of a State Party in the Hague and the desk(s) in the foreign ministry or National Authority in the State Party but not the public. Implementing the CWC means more than doing exact bureaucratic and legalistic paper work.

Of course, exact administrative and legal action is an important precondition for success. However, disarmament law is a project to change society. This change will take place only when it is generally promoted. A new reality will exist when all CW have been abolished and their reappearance is impossible because of the civilising impact of the Convention on human society. Without the support of 'humankind' we cannot honestly assume that we will be successful in the long run. Public opinion condemns poisonous weapons as especially heinous; the world-wide uproar about the chemical massacre in Halabja (Iraq) had their share in the successful negotiation process of the CWC. A special programme is needed to win the support of people in all countries to implement the CW ban. The title of such a programme should be "Freedom of Information Programme of the OPCW". It would provide the necessary basis that the OPCW becomes publicly accountable for its work.

3. CONCLUSIONS

"An asset for the sake of peace and humanity - 10th anniversary of the Chemical Weapons' Ban" is the title of an appeal launched on 13 January 2003. Up until now, it has been signed by personalities and institution that actively participated in the negotiation and implementation of the CWC [14].

In the Appeal the signatories appreciate the conclusion of the CWC and the results of its implementation achieved. At the same time they call for increased efforts for achieving universality, to promulgate national implementation legislation, to assist the Russian Federation's CW destruction program. They call on the political organs to make decisions consistent with the requirements of the Convention, to respect the rights of inspectors and the independence of the Technical Secretariats' staff from instruction of States Parties.

In the light of the results of the First Review Conference of the OPCW the conclusions of the Appeal are still up-to-date. Only the request to scrap the decision of the Conference of the States Parties for a tenure policy of 7 years maximum for staff including inspectors has to be added.

The conclusions of the Appeal read:

"Failure to resolve problems, such as those illustrated above, will, inevitably, seriously impede the implementation of the Convention and will have serious consequences for the prevention of proliferation and the appearance of new chemical weapons. At a time in which security against any use of weapons of mass-destruction is of vital importance it is essential that the potential of the Organisation for the Prohibition of Chemical Weapons to verify the disarmament and non-proliferation of chemical weapons be used to the fullest extent.

All States need to contribute to the strengthening and the effective universality of the Convention. We, therefore, appeal to the States Parties to the Chemical Weapons Convention to give due consideration to the following:

Strict observance of the Chemical Weapons Convention is an essential component of international peace and security. Regrettably, governments have reduced, to less than a routine level, the attention they give to the Convention. A proactive policy is needed, geared to the full implementation of the Convention and its adaptation, where appropriate, in the light of experience gained during its first five years of implementation, scientific and technological advances and the new challenges posed by the threat of chemical terrorism. Success is most likely to be achieved by the encouragement of full compliance with the comprehensive prohibitions of chemical weapons, the Convention's requirements by all States Parties and also by the Organisation moving towards a much greater level of transparency in relation to compliance related issues.

The way ahead calls for:

- a change in the present restrictive attitude, by some States Parties, towards the CW Convention and its Organisation and an effort by all States Parties to redress the damage inflicted on it;
- a resolute effort by the political organs of the Organisation to concentrate on issues of compliance and to inform the public about those issues - they must be prepared to take difficult decisions more effectively and more transparently; and

- a return to the basic consensus developed during the negotiation of this Convention: each State Party must be convinced by objective and impartial procedures that all other States Parties fully comply with and abide by their respective obligations.

Looming dangers of war should inspire such action to maintain the purposes and objectives of the Convention."

References

[1] OPCW (2003) Political Declaration, available at: www.opcw.org/cwrevcon/doc/NAT/FRCpoliticaldeclation.html. OPCW, The Hague.
[2] see, for instance, International Labour Organisation Judgement 102232 (re Bustani)
[3] In fact, the IAEA has taken the CWC as a precedent for attempting to expand the scope of the privileges and immunities accorded to the Agency. See: the Board of Governors document GOV/INF/680 of 10 February 1993.
[4] Krutzsch, W., Trapp, R. (1999). Verification Practice under the Chemical Weapons Convention, Commentary, Kluwer Law International, The Hague/London/Boston, p.20.
[5] Woolomes Tabassi, L. (ed) (1999) OPCW: LEGAL TEXTS, T.M.C.Asser Press, The Hague p.188. The same decision infringes also upon other issues concerning the immunity of inspectors equipment: Subparagraph 3.2.4 of this decision claims a right of the representative of the inspected State Party, at the stage when the preliminary findings have been prepared, to have information which he believes is not related to the purpose of the inspection removed in his presence from the recording media. This is not consistent with paragraph 11 (c) nor with paragraph 60 of Part II of the Verification Annex. According to subparagraph 3.2.5, an inspected State Party might even, in an exceptional case, when the removal procedure cannot be followed (to be decided by this State Party) retain and confiscate the recording media of the inspectors.
[6] Sunshine-project (2002) available at: http://www.sunshine-project.org/publications/pr240902map.html.
[7] The CBW Conventions Bulletin, Quarterly Journal of the Harvard Sussex Program on CBW Armament and Arms Limitation, issue NO 61 September (2003) 'Non-Lethal' Weapons, the CWC and the BWC. pp. 1-2.
[8] Krutzsch, W., (2003) Non-lethal chemicals for law enforcement? available at: http://www.bits.de
[9] Vertic (2002). Getting verification right (The report pleads for "scrapping of the rigid fixed tenure policy).
[10] A sophisticated experts draft exists unused since years – see Krutzsch/Trapp, ibid. Footnote 274, p. 87.
[11] This is a non-issue, see: Krutzsch, W., Trapp. R., ibid., Part VII (B), paragraph 7.

[12] For example, the detailed plans for verification of the destruction of Category 2 chemical weapons at Shchuch'ye in the Russian Federation first came up before the Council in October 2000 (at its twenty-first session). It was subsequently considered and decision deferred by the Council at each session until the twenty-seventh sessions and the sixteenth meeting of the Council in November 2001. The issue of contention was the draining and transport of the Schedule 3 chemical phosgene from munitions stored at Shchuch'ye. At the Council's twenty-eighth meeting in March 2002, it was brought to the Council's attention that the draining of the phosgene and its destruction had been completed in the absence of a Council decision (with the presence of OPCW inspectors); moreover, the destruction had taken place at a facility and in a manner objected to by the majority of Council members. At this point, all the Council could do was state that it does not consider the Russian actions as establishing a precedent. A similar trend of deferred decisions has been witnessed with respected to the combined plans for destruction and verification of chemical weapons production facilities, and to requests for conversion of chemical weapons production facilities.

[13] Paragraph 29 of Article VIII of the Convention stipulates two-thirds majority of all its members for decisions on substance and simple majority for decisions an procedural matters.

[14] Appeal (2003), An asset for the sake of peace and humanity - 10th anniversary of the Chemical Weapons Ban, available at: http://www.cwc-support.

SESSION I: PREDICTION

EVALUATION OF STATE OF ENVIRONMENT AND MONITORING OF HAZARDOUS FACILITIES

Mikhail Kurguzkin[1]
[1]*Ministry of Natural Resources and Environmental Protection of the Udmurt Republic, Izhevsk (Russia)*

Abstract: Safe operation of high hazard facilities located within the boundaries of the regions and republics of the Russian Federation can be achieved by joint work of several government bodies. Each of them is to efficiently execute its specific functions. However, effective coordination of their work is a specific problem. Coordinated work depends very much on the information about the hazardous facilities and their respective areas. This is why the role of each government body should be clearly determined on all stages of joint work: collection of information, its analysis, and decision-making. The presented report is dedicated to creation of a system of environmental monitoring on the territory of one of the Russian Federation regions.

Key words: high hazard facility, environmental monitoring, state of environment, operational control, database system

The Udmurt Republic is one of few regions of Russia where high hazard facilities are more numerous in comparison with other regions. Most importantly, this concerns chemical weapons stockpiles and facilities for disposal of military equipment.

From the ecological point of view, the term 'high hazard facility' is related to risk of emergency situations, which can result in releases of considerable amounts of particularly dangerous (toxic) substances into the environment or emergency operation of the facility without observation of prescribed environmental standards, which leads to the same result over a longer time period.

Since ensuring the safety at such facilities is the responsibility of several government bodies serving various purposes, effective coordination of their activities is impossible without clear understanding of their respective places in the safety system, and of how they interact, especially at informal level..

The existence of the problem of interaction is evident. It can be solved only with the initial determination of the functions of the participants of the safety ensuring process.

V.M. Kolodkin and W. Ruck (eds.),
Ecological Risks Associated with the Destruction of Chemical Weapons, 41–44.

In this connection analysis of certain aspects of activities involving the monitoring of the high hazard facilities and their areas is of interest.

It is generally accepted that the notion of 'monitoring' comprises three stages: acquisition of information (for example, by direct observation and measuring), its analysis and prediction of further development of the situation or process. These three components are typical both for the monitoring aimed at prevention of emergency situations and for a so-called 'environmental monitoring' aimed at minimization of the negative after-effects to the environment from hazardous facilities. Though both goals look similar, the methods of their achievement are quite different. What are their basic differences?

First of all, they are in the observed objects. In the first case, we deal with a source of probable emergency situation, for example, chemical agents storage, process area for disassembling chemical munitions, incinerator for solid rocket fuel, etc. The objects of the environmental monitoring are elements of animate and inanimate nature in specially selected monitoring sites. The concentrations of pollutants of the free air and water sources are determined by the hydrometeorological service, and the contamination of soil and ground deposits, as well as the general state of flora and fauna, are to be evaluated by the government bodies responsible for the environmental protection.

Quite close to the notion of 'environmental' is the monitoring of biological communities and changes in the diversity and number of their species. At the same time, as practice shows, the connection of these changes with the influence of the facility polluter is not always clear and cannot be expressed quantitatively. However, this type of observation is desirable, especially when the existence of this influence is already known and proven.

The levels of environmental pollution from the point of view of influence on the state of health of the population should be the objects of monitoring to be carried out by the State Sanitary-Epidemiological Inspection. Since the hazardous facilities raise high concern of the residential population, the public should have access to up-to-date reliable information about the sanitary-epidemiological situation in their residential areas.

Monitoring, i.e., control of sources of hazard, in contrast to the above described observations, does not require any fine quantitative analyses, as its goal is to get a signal on the operational measures to be taken to prevent an emergency or to eliminate it. For this aim, in most cases, a simple registering equipment with 'threshold' actuation can be used.

The time scales determine the difference between the 'operational' monitoring and 'environmental' monitoring. In the first case, according to the determined goal, the sequence 'observation-analysis-prediction-response' is realized in real time and over a minimum possible time period. As to the accumulation of pollutants and changes in the structures of biological communities under the influence of the hazardous facility, it can go on for months, even years.

Based on the above it is possible to state that environmental monitoring is a specific insurance against both risk and factors of negative effect not considered within the assessment of the influence on the environment, or by the environmental expertise which serves as an instrument for revealing the above factors.

Another important function of the environmental monitoring is revealing changes in the environment after the end of operation of the hazardous facility with the aim of determining the possible damage.

The results obtained by the operational control technique can be applied to the needs of monitoring, but the reverse use is not always possible. For instance, the complete volume of information acquired by the environmental monitoring is excessive and useless for the practical needs of the Ministry for Emergencies. The important difference between the environmental monitoring and operational control is that the fromer, unlike the latter, deals with the negative response of the animate and inanimate elements of nature.

As to the survey of areas for chemical or radiation contamination after an accident, disaster or an emergency, the aim of this activity is, most importantly, to obtain information to keep people away from the dangerous area and to take the relevant measures within the dangerous area. In this case we can talk about minimization of the consequences of the emergency. Further environmental monitoring is also necessary, but it will follow other goals and should be organized in a special manner in contrast to monitoring at the normal operation of the facility.

Thus, a very important conclusion can be made that the organization of environmental monitoring is a more complicated task in comparison with the operational control over the high hazard facilities and elements of their structure. This difference, as follows from the above, is due to the complexity of the environmental monitoring and observed processes. Besides, fundamental difficulties arise when it is necessary to analyze heterogeneous information with the aim of building integral assessments of the quality of the environment.

Organization of the environmental monitoring relative to the high hazard facilities requires working-out a special program particularly taking into account the characteristics of the facility and their impact on the environment. The bases of such programs at the stage of creation of the design documentation are assessments of the effect on the environment. In principle, the same information is necessary to develop programs of environmental monitoring for functioning facilities, in this case use is made of real and not of predicted effects on the environment.

All the above considerations show the basic differences between what is called environmental monitoring and the activities on control over the high hazard facility. Based on the above, it is possible to formulate a range of principally important and first-rank priorities on the development and interaction of various government bodies and their observation systems on the territory of Udmurtia:

1. Development of material resources and the professional community of the Agencies responsible for monitoring systems.
2. Upgrade of up-to-date informational technologies (GIS, communications means).
3. Standardization of software for all observation systems.
4. Coordination of the necessary nomenclature, volumes and periods of information exchange and information transfer protocols.

The solution of these problems will allow passing to the next stage of development and interaction of the control and monitoring systems. Evidently, all information flows should be combined in one common Information Analytical Center. One variant can be the creation of this center based on the Ministry for Emergencies.

PREDICTION OF QUANTITATIVE ASSESSMENTS OF EFFECTS ON NATURE FROM POTENTIAL ACCIDENTS AT CHEMICAL WARFARE AGENT FACILITIES

Vladimir Kolodkin[1], Aleksey Murin[1]
[1]*Research Institute of Natural and Technogenic Disasters, Udmurtia State University, Izhevsk (Russia)*

Abstract: Presented is a technique of prediction of quantitative risk assessments for Nature from potential accidents at chemical warfare agent facilities.

Key words: environmental risk, accident, chemical weapons

1. INTRODUCTION

One of the key problems in the complex of problems associated with chemical weapons' storage and destruction is the problem of insuring safety. This problem cannot be solved without the preliminary analysis of all the possible aspects of the potential damage. It is possible to suppose that the level of hazard associated with chemical weapon storage and destruction won't exceed the designed level on condition of trouble-free operation of the facility. Analysis of the accident hazard that may occur as a result of unauthorized release of chemical agents seems more important. The causes leading to the accident at the facility with warfare chemical agents can be diverse. This can be such a low-probability event like the fall of a meteorite or an aircraft crash over the stockpile site. This can be the result of illegal actions (terrorist attack) or a natural disaster. It is not possible to exclude from consideration errors of personnel that are probable during work at the facility.

The quantitative characteristics of the level of the potential accident hazard is expressed by accident risk assessments. The prediction technique of accident risk assessments relative to the human being is developed nowadays to the level of prediction of quantitative assessments [1-3]. The existing discrepancies in the numerical values of accident risk assessments done for the human being [1-3] are due not to the differences in the techniques used but, first of all, to the uncertainties in the description of

V.M. Kolodkin and W. Ruck (eds.),
Ecological Risks Associated with the Destruction of Chemical Weapons, 45–55.

accident processes and, as a result, to the difference of the physical-mathematical models used in prediction.

However Russia has no prediction technique for risk assessments relative to Nature which would be generally accepted by the scientific community. Some works on prediction of risk assessments for Nature [4-6] deal with continuous influence of technogenic facilities on the environment and do not touch upon the problems of prediction of accident risk assessments. This contribution presents a technique for prediction of quantitative risk assessments for Nature describing the after-effects of potential accidents at chemical warfare agent facilities.

2. RISK ASSESSMENTS FOR NATURE

In analysis of the level of the potential accident hazard for Nature the number of risk recipients is important. In the general case the term "the risk recipients of Nature" means elements of the animate nature (biota): human beings, fauna (ground and aquatic) and flora (ground and aquatic). The quantitative risk assessment should be computed relative to each risk recipient. This is why, if the quantitative risk assessment is considered relative to Nature, this risk can be presented by a vector, each component of which is a characteristic of the level of hazard relative to a specific risk recipient. Generally, this vector value fully characterizes the level of hazard inflicted by the chemical agent facility.

All the body of mathematics of the theory of accident risk [1] is generalized for risk assessment relative to Nature. Let us introduce the following notation: P_z is the probability of the z-th accident process, $(U_k^z)_i$ is the predicted damage from the z-th accident process for the i-th (i = 1,2,3, …, ∞) risk recipient. In the adopted notations the expression for the i-th component of the accident risk R_i characterizing the level of hazard relative to the i-th recipient of risk at a certain point (r, φ, h,t) can be presented as:

$$R_i(r,\varphi,h,t) = \sum_z P_z \cdot W^z{}_i(r,\varphi,h,t), \qquad (1.1)$$

where the relative accident risk assessment associated with the z-th accident is

$$W^z{}_i(r,\varphi,h,t) = \sum_k \nu_k \cdot (U_k^z)_i(r,\varphi,h,t). \qquad (1.2)$$

The relative accident risk assessment characterizes the predicted after-effects of the accident of the specified type. The expression for the relative accident risk assessment comprises the probability of occurence of the k-th set of climatic characteristics of the environment ν_k. The relative accident risk assessment W^z_i is a characteristic of the level of hazard inflicted by the technogenic facility on the i-th risk recipient on condition that the accident situation developed through a specified scenario. Specific expressions for relative accident risk assessment are built depending on the type of the accident effect and way of its transmission. In the case under consideration there is introduced a polar coordinate system (r,φ,h) with the beginning at the point of the center of the source of hazard; the time t being calculated from the moment of accident.

The damage function $(U^z_k)_i$ expresses the probability of damage to the i-th risk recipient. In particular, the possibility of lethal outcome of the accidents associated with release of highly toxic substances can be considered as a damage function relative to a fauna representative; the loss of the respective representative relative to a flora representative. The shape of the damage function for the most significant risk recipients is found in special experiments. In this particular case, the damage value is determined by the value of the toxic dose D(r, φ, h, t) to which the risk recipient is exposed.

For example, the damage function for the human being at the inhalation exposure is

$$U(D)=1/[1+(LC_{\tau 50}/D)^{\beta}], \qquad (1.3)$$

$$\beta=1{,}677/\ln S, \qquad \ln S=0.336.$$

The toxic properties of the chemical agent are characterized relative to the risk recipient at the inhalation exposure by the value $LC_{\tau 50}$ which is the average lethal dose (mg*min/m^3) and the toxicity function S determining the resistance of the organism to the effect. The numerical value $ln\ S$ is determined by the experimental data.

The toxic load at a space point (r, φ, h) by the time moment t under the given climatic condition and at a specific potential accident scenario is normally subject to computation. In this case we arrive at the assessment of the area risk which is responsible for the predicted probability of affliction of the i-th risk recipient at a specific point. This being the case, it is assumed that the risk recipient stays at the specified point (r, φ, h) until the time moment t without any protective means. That is, the area risk assessment is an inherent characteristic of the space point relative to the level of hazard for the i-th risk recipient.

The dissection of the space field of the area accident risk for the i-th recipient by the plane h=const can be shown on the terrain map. In doing so, the zones responsible for various risk level R^*_i are delineated. The zones are delineated by the condition $R_i(r, \varphi, h^*, t) > R^*_i$. The zone area is determined according to the map of the facility site. If several risk levels R^*_i are brought into consideration, the area of affliction of the i-th risk recipient is differentiated by the levels of the potential hazard. The area dimension is characterized by the effective radius $(R_{eff})_i$.

The relative group accident risk assessment at a point $(r, \varphi, h^*) \in \Omega$ is

$$(W^z_g)_i(r,\varphi,h^*,t) = M_i\ \psi_i(r,\varphi,h^*)\ W^z_i(r,\varphi,h^*,t), \tag{1.4}$$

where $\psi_i(r,\varphi,h^*)$ is the density of distribution of the i-th recipient over the afflicted zone Ω, $(r,\varphi,h^*) \in \Omega$ corresponding to the normality condition $1 = \int \psi_i(r,\varphi,t^*)d\Omega$; M_i is the total number of the i-th recipients in the area Ω.The group (population) risk assessment characterizes the level of the potential risk at a point (r, φ, h^*) taking into account the distribution of risk recipients over the afflicted zone Ω. By its definition the assessment corresponds to the probability of affliction of the risk recipients by the z-th accident at a point (r, φ, h^*) by the time moment t.

The level of the accident hazard associated with a specific accident and localized in a certain area Ω by the time moment t is characterized by the integral relative accident risk assessment $(W^z_L)_i(\Omega, t)$ and the integral relative group accident risk assessment $(W^z_G)_i(\Omega, t)$:

$$(W^z_L)_i(\Omega,t) = \int_\Omega W^z{}_i(r,\varphi,h^*,t)d\Omega\,. \tag{1.5}$$

$$(W^z_G)_i(\Omega,t) = \int_\Omega (W^z_g)_i(r,\varphi,h^*,t)d\Omega\,. \tag{1.6}$$

In the particular case the integration area $(r, \varphi, h^*) \in \Omega$ can include all the afflicted area.

The integral relative accident risk assessment $W^z_L(\Omega,t)$ characterizes the predicted after-effects of a specific accident in the area Ω. Its value does not depend on the population distribution over the afflicted area Ω and is determined by the inherent properties of the facility and accident. The integral relative accident risk assessment $W^z_L(\Omega,t)$ can be interpreted as the zone area of the 'certain' affliction of the risk recipient, i.e., the zone where the probability of affliction of the risk recipient is close to unity. In this regard, the assessment is normally used for comparison of the after-effects of the accidents of various types and for comparison of various technogenic facilities relative to the potential hazard inflicted by the facilities.

In essence the integral relative group accident risk assessment corresponds to the predicted number of risk recipients afflicted by a specific accident in the area delineated by a specified sign. For example, the integral relative group accident risk assessment can characterize the level of the potential hazard from a specific accident within a populated area, reserve, etc. This value can be interpreted as a predicted number of victims within a delineated zone Ω by the time moment t at a specific accident at the facility.

Summing the relative risk assessments with the corresponding weighting factors transforms the relative risk assessments W_i into the risk assessments R_i in accordance with relationship (1.1). Summation covers the accident risk assessments of all possible accidents or accidents of a certain type. Practical computation considers only the most significant accidents (by their after-effects).

3. PREDICTION OF RISK ASSESSMENTS FOR NATURE

Prediction of risk assessments includes the following stages:

1. Building scenarios of accident situations and their frequency analysis. This stage requires information on production processes, retrospective data, etc. As a result, sets of scenarios and their probabilities are built.
2. Mathematical modeling of sources of hazard. The stage of emission of a hazardous substance (mass, energy) into the environment is modeled for each scenario. Modeled are the most hazardous scenarios. The informational provision of this stage requires the existence of information on the physical-chemical properties of hazardous substances and compounds. The results of Stage II are used as the initial conditions for the implementation of further computations.
3. Mathematical modeling of the propagation of the accident effects in the environment. Modeled are several accident scenarios at several probable states of the environment. The information used in computation is about the character of the underlying surface (relief, structure of land tenure), climatic data (long-term temperature distributions, wind speed and direction, precipitation, etc). Computation of one case with the use of an express model takes a few minutes. Computation by the three-dimensional model, which gives the completed picture of the event, takes up to several hours of operation of a highly productive cluster comprising 10 processors of the Intel Pentium III 700 MHz type. The result of the

stage is a database with the space-time distributions of afflicting factors (with a base volume of about 1 GB).

4. Prediction of risks. The implementation of the stage comprises computation of the space-time structure of area and group risks, topologic characteristics of risk fields, structure of the significance of various accident scenarios and conditions for their realization. Computed are the expected values of the damage function on the totality of occurences. About 100 accident scenarios and about 100 probable states of the environment are modeled in predicting risk assessments. Risk computation requires hundreds of hours of operation of a multi-processor cluster. The operation results in detailed information on the distribution of risks over the real time interval of 3 hours from the moment of the accident. The information provision of the stage includes: databases on the characteristics of accident effects, space distribution of the number of risk recipients, characteristics of afflicted recipients.
5. Risk analysis.

4. ACCIDENT RISK ASSESSMENTS FOR NATURE ASSOCIATED WITH CHEMICAL WARFARE AGENTS STOCKPILE

Computation of quantitative characteristics of the accident risk for the chemical weapon facilities relative to Nature requires, in the general case, considerable informational and computational resources. This is why, usually, researchers restrict themselves in consideration of accident scenarios to the analysis of the accident after-effects which make the greatest contribution to accident risk assessments: analysis of after-effects of 'additional to the plan' accidents. This leads to the upper bound of the potential accident hazard.

The second factor that permits considerable decrease in the demands for resources is consideration of the specific character of action of chemical warfare agents. In particular, if we restrict ourselves to the analysis of the accident after-effects leading to unauthorized release of organo-phosphorus warfare agents, it is necessary to note that the effect of nerve chemical agents (sarin, soman, VX) with anti-choline-esterase effect is important only for mammals and humans. This means that many risk recipients can be reduced to a subset of mammals and humans.

Table 1 presents the characteristics of the toxic effect of chemical warfare agents on the human being and some representatives of mammals.

The characteristics of the toxic properties belong to the inhalation pathway for an exposure duration up to one hour.

The analysis of experimental data shows that the response to the toxic effect of sarin and soman is proportional to the mass of the risk recipient. The literature data gives:

Sarin $$Ln[LC_{\tau 50}/(LC_{\tau 50})^{*}] = 0.38 - 0.16 * Ln(P) \quad (4.1)$$

Soman $$Ln[LC_{\tau 50}/(LC_{\tau 50})^{*}] = 1.24 - 0.41 * Ln(P) \quad (4.2)$$

where $(LC_{\tau 50})^{*}$ is the lethal dose for the human being, $LC_{\tau 50}$ is the lethal dose for the mammal, P is the mammal weight.

Table 1: Characteristics of toxic effect of chemical warfare agents on animals and human being

	Sarin		Soman		Vx		Lewisite	
Species	LD_{50} mg/kg (skin-res.)	LC_{t50} mg*min/m^3 (inhalation)	LD_{50} mg/kg (skin-res.)	LC_{t50} mg*min/m^3 (inhalation)	LD_{50} mg/kg (skin-res.)	LC_{t50} mg*min/m^3 (inhalation)	LD_{50} mg/kg (skin-res.)	LC_{t50} mg*min/m^3 (inhalation)
Mouse	33[11]	240-310[11]	2[11]		0.046-0.092[11]	4.0[11]	15-17[11]	494
Rat	48.5[11]	80-220[11]	3[11]	4.5[11]	0.065-0.1[11]	17[11]	24[11]	585
Guinea-pig	18.1[11]	74[11]			0.035[11]	8.3[11]	12[11]	
Rabbit	20-22[11]	120[11]	0.8[11]		0.013-0.025[11]	25[11]	6-7[11]	5694
Dog	15-40[11]	60-80[11]	3.3[11]	26.0[11]	0.015; 0.054[11]	15[11]	15[11]	650
Pig	116[11]	34[11]			0.4[11]			
Monkey		74[11]			0.06[11]	50[11]		
Cat					0.02[11]		10-12[11]	
Human being	24-28[11] 26[11]	75[9]- 70-100 [8,11,14] 70[15]	4.0 - 10[11]	30[9]- 50[11] 70[16]	0.1[9]	10[9], 32[11] 35-36[11]	80[11]	1500 [7,12,13], 1200-1300[11]

The risk recipient response to VX does not depend on the risk recipient's body mass; in the first approximation the response can be considered constant for all risk recipients.

The above assumptions are satisfactory for prediction of accident risk assessments. Some results are presented in Fig. 1 showing the positions of the lines of the equal values of the area accident risk for mammals (mice, rabbits) and humans for an accident at the storage arsenal for artillery shells filled with sarin, soman and VX. Table 2 shows the effective radii of the zones of equal values of the area risk.

Table 2: Effective radii of zones corresponding to risk R at accident at Kizner Arsenal

Risk recipient	Radius of zone (m) with risk, minimum: $R=10^{-7}$	Radius of zone (m) with risk, minimum: $R=10^{-6.5}$	Radius of zone (m) with risk, minimum: $R=10^{-6}$	Radius of zone (m) with risk, minimum: $R=10^{-5}$
Mouse	1918	1440	1034	529
Rabbit	4627	2920	1542	946
Human being	8660	5905	3140	2163

The quantitative assessments show that the human being is the most vulnerable risk recipient at the accident. When the effective radius of the safety zone for the human being is 1918 m, for mice it is 529 m.

Fig. 1: Isolines of area risk for Kizner Chemical Warfare Agent Stockpile site

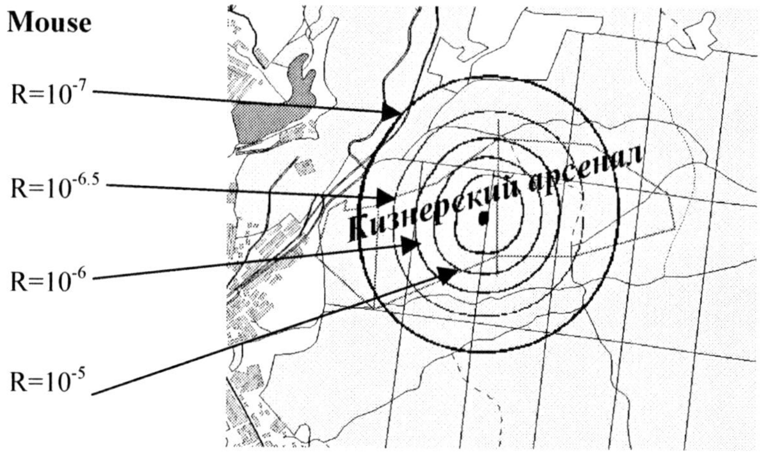

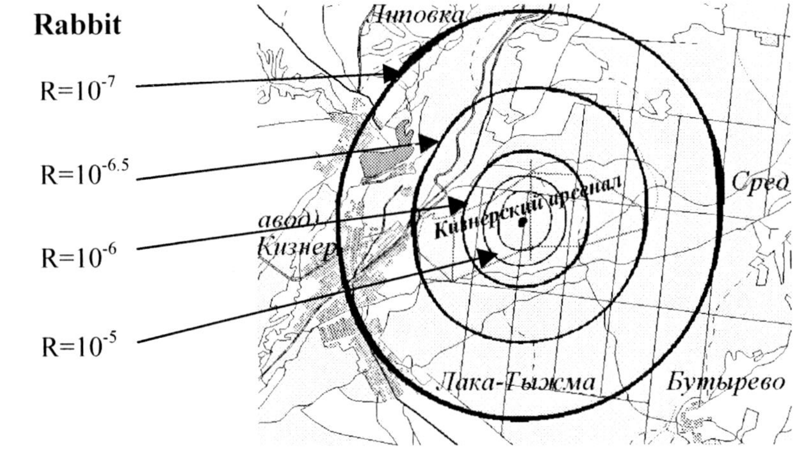

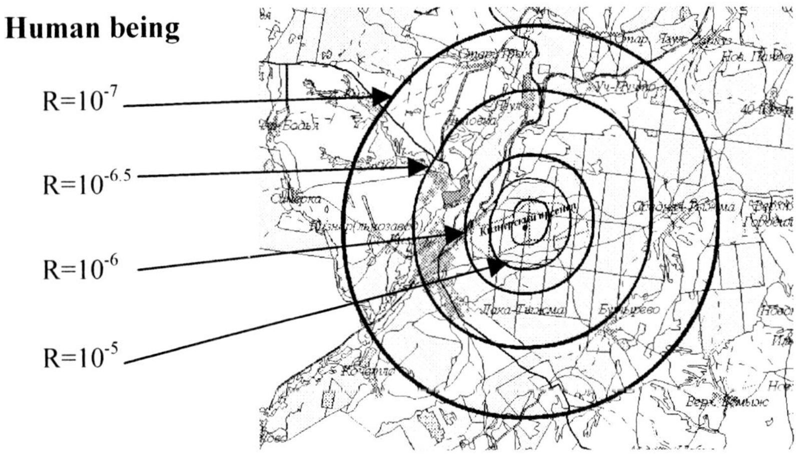

5. CONCLUSION

The accident risk assessment for Nature can be characterized by a vector value which each component characterizes the level of hazard for a specific risk recipient. Accidents at chemical weapon facilities leading to release of organo-phosphorus agents into the atmosphere are more hazardous for humans.

ACKNOWLEDGMENT

The work was made possible thanks to the financial support of the Russian Foundation of Basic Research (Grant 01-01-96444), Research Programme "Russian Universities" (Grant УР.03.01.015) and the ISTC foundation (Grant # 2065).

References

[1] Kolodkin, V.M., et al. (2001) Quantitative Risk Assessment of Chemical Accidents. Izhevsk, Udm. Univ. Publ.House.

[2] Declaration of Safety of Facility for Chemical Weapon Destruction in Shtchuchye District, Kurgan Oblast on Stage of TEG. (1999) M.

[3] Russian Complex for Chemical Weapon Destruction. Risk Assessment for Population. Report No. SAIC-01/2607 (2002).

[4] Method of Determination of Prevented Ecological Damage. State Committee of RF for Environmental Protection. M (1999).

[5] Murzin, N.V., Bykov, A.A. (2003) Calculation of Damage for Natural Communities and Risk for Ecosystems from Effects of Poisonous and Toxic Substances. Problems of Analysis and Risk Management. V. 1., 1: 32-50.

[6] Murzin, N.V., Lystsov, V.N. (2003) Analysis of Effect of Poisonous and Toxic Substances on Ecosystems. Problems of Analysis and Risk Management. V. 1., 1: 51-78.

[7] Carlsson, E., Konberg, M., Runn, P. (1995) Anaiysis of After-effects of Possible Accidents at Lewisite Stockplie in Kambarka Area. Russian Chemical Journal 4.

[8] Franke, Z. (1973) Chemistry of Poisonous Substances. M., Chemistry.

[9] Aleksandrov, V.N., Emelyanov, V.I. (1990) Poisonous Substances. M. Voenizdat.

[10] Recommendations on Organization of Protection of Population Living Near Chemical Warfare Agent Storing and Destruction Facilities and Interaction of Government Bodies in Emergencies at These Facilities. (1996) M. VNIIGOCHS

[11] Risk Assessment Associated with Chemical Warfare Agent Stockpiles on Territory of Udmurtia Republic.(1996) Ed. by Klodkin, V.M. Izhevsk, UdSU Publ.House.

[12] Military Chemistry and Chemical Compounds. Army Air Force. (1975) .P. 11-12

[13] Military Chemistry and Chemical Agents. Departments of the army and the air force.(1963) P.26-27

[14] James, A. F. Compton. Military Chemical and Biological Agents. The Telford Press Caldwell, NJ

[15] http://www.cbdcom.apgea.army.mil/RDA/erdec/risk/safety/msds/gb1.html

[16] http://www.cbdcom.apgea.army.mil/RDA/erdec/risk/safety/msds/gd1.html

ECOLOGICAL RISKS OF TOXIC SUBSTANCES COMBUSTION

Alexey M. Lipanov[1], Mikhail A. Korepanov[1], Zufar A. Tukhvatullin[2]
[1]*Institute of Applied Mechanics, Ural Branch Russian Academy of Sciences, Izhevsk (Russia)*
[2]*State Corporation «Votkinskiy zavod», Votkinsk (Russia)*

Abstract: Experimental and theoretical results of the authors and other Russian and foreign scientists on analysis of toxic substances formation from combustion of chemical compounds are examined. The theoretical approach of the authors has a general character, allowing to determine concentration limits of substances formed by high temperature combustion and can be applied to a wide set of utilized toxic substances. Special attention is given to the case of combustion of chlorine-containing substances, during which formation of polychlorinated aromatic hydrocarbons (dibenzo-*p*-dioxines, dibenzo-furans, etc.) is possible. The report shows that a correctly organized combustion process, with fast water cooling of combustion products in particular, the contents of dioxins and furans in combustion products can be reduced 8-10 times.

Key words: theoretical approach, experimental research, combustion, dioxine

1. INTRODUCTION

During its industrial activity mankind has created super-toxic substances like polychlorinated dibenzo-*p*-dioxines, dibenzo-furans, representing special danger for people and the environment [1]. In general there are 210 such substances, all of them differing in their physical, chemical and toxicological properties, the most toxic being 2,3,7,8-tetra-chlorodibenzo-*p*-dioxine (2,3,7,8-TCDD). Doxins surpass most pollutants by toxic level: the maximum concentration limit of dioxins in air converting to 2,3,7,8-TCDD is 0.5 pg/m^3, whereas for carbon monoxide it is – 3 mg/m^3.

V.M. Kolodkin and W. Ruck (eds.),
Ecological Risks Associated with the Destruction of Chemical Weapons, 57–63.

2. RESULTS AND DISCUSSION

The most commonly used method to determine the possibility and state of formation in thermal processes of different substances is a thermodynamic calculation based on chemical equilibrium. This can be explained by the lack of data on chemical reaction kinetics of complicated organic substance formation. For example [2, 3] can be mentioned, where the possibility and condition of dioxins formation in thermal processes in sinter plant and solid waste incineration are described. In [2] researches of the influence on dioxin and furan formation process, of temperature, free oxygen concentration, chlorine content, carbon, hydrogen in exhaust gas, and the correlation between hydrogen and carbon, are cited in detail. Results of numerical calculation are compared with experimental researches, in which measured partial pressure of dioxins and furans in exhaust gases of sinter plant were approximately 10^{-13} and 10^{-11} respectively.

For the theoretical determination of the composition of high temperature gaseous mixtures the data of thermodynamic properties (heat capacity, enthalpy, entropy) of substances is sufficient. In this work data of [4-8] was used.

Fig. 1. Partial pressure of dioxin and furan

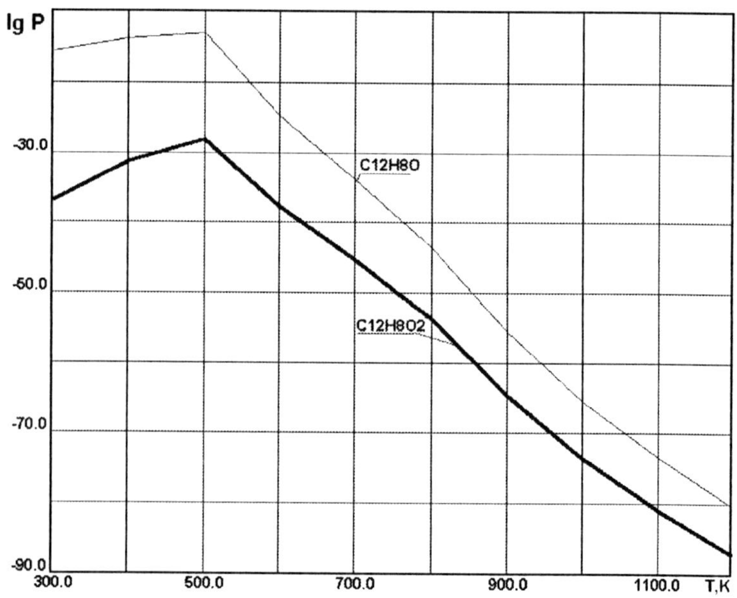

In Fig. 1 the results of a test calculation for a gaseous mixture with element ratio C:H:Cl:O = 1:1:1:0.5 are shown. The results are in accordance with results from [2, 3]. Fig. 1 shows that the concentration of dioxins and furans have a maximum between temperatures 400-750 K equilibrium. As result of test calculation, it was found that by ratio C/H > 2, exclusion of soot from the calculation is necessary to avoid its influence on dioxine concentration that corresponds to [2].

One of the examples of complicated chemical compound destruction is the combustion of solid propellant based on ammonium perchlorate, NH_4ClO_4. It is considered [9] that burning large quantities of polychlorinated dibenzo-*p*-dioxines and dibenzo-furans up to 10^{-10} kg/kg of propellant are formed in solid propellants..

For experimental research purpose a plant was created (Fig. 2). Solid propellant (1) is placed inside of the reactive engine model (2, 3). A gas flue is connected to the engine, along which over 330 mm 3 rows of sprayers (6) are placed to add cool water. A reservoir (10) with volume 2 m^3 is connected to the gas flue. In the upper part of the reservoir a sprayer (12) is mounted for washing the inner wall of the reservoir. To add cooling water a high pressure reservoir (15) with volume 0.3 m^3 is used.

For prediction of process parameters, a theoretical calculation using chemical equilibrium was made. This calculation allows prediction of the maximum possible quantity of toxic substances formed in the combustion process. In Fig. 3 the quantity of toxic substances, as mass fraction in combustion gases of solid propellant without chemical interaction with cooling water, are shown. The maximum quantity of hydrocarbons, among them polychlorinates, is reached at a temperature between 600-1200 K. Taking into account chemical interaction of solid propellant combustion gases with cooling water, the mass fraction of toxic substances decreases 8-10 times. This can be explained by the decrease of C/H ratio, which, in correlation with [2], leads to a decrease of dioxin and furan formation. However, taking into account that vaporization of cooling water leads to a fast temperature decrease in combustion gases, it is possible to not consider water vapor in chemical reaction.

Fig. 2. Scheme of experimental plant

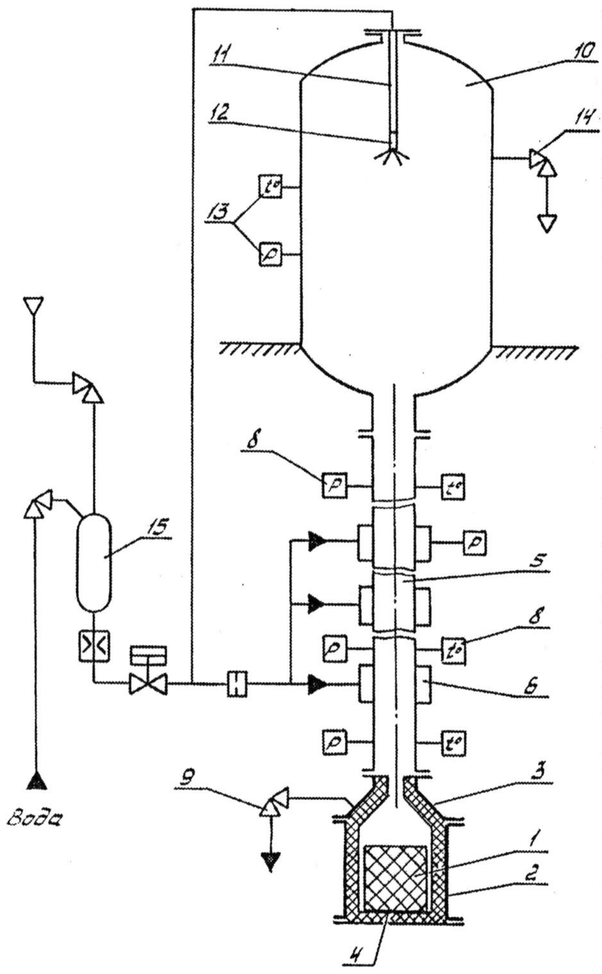

Fig. 3. Mass fraction of dioxines and furans depending on temperature

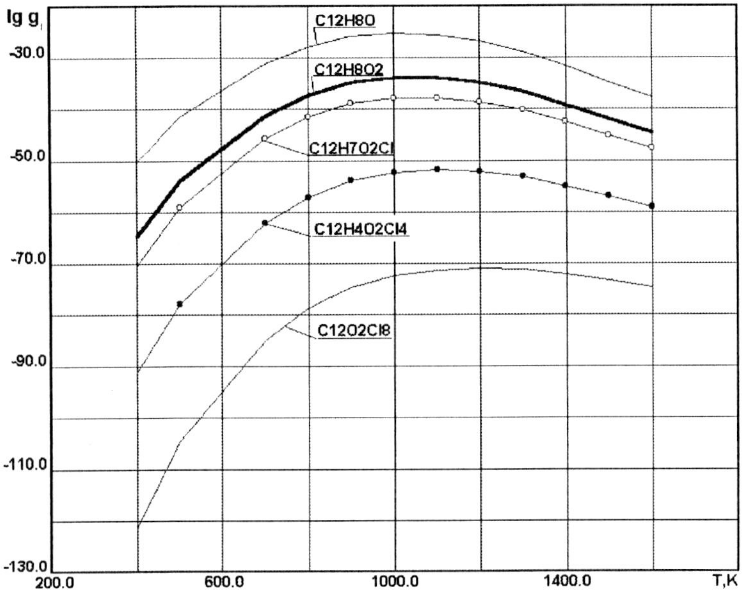

Table 1. Content of pollutants in combustion gases of solid propellant (mass fraction)

No.	Substance	Solid propellant		Experim. data [9]
		theor.	exper.	
1.	Benzol	$1.52 \cdot 10^{-11}$	NA	-
2.	Phenol	$2.19 \cdot 10^{-14}$	NA	-
3.	Biphenyl	$8.60 \cdot 10^{-23}$	NA	-
4.	Polychlorinated biphenyls	NA	NA	$52.2 \cdot 10^{-9}$
5.	Dibenzo-furan	$6.53 \cdot 10^{-26}$	NA	-
6.	1,2,3,7,8- chlorodibenzo-furan	NA	$0.58 \cdot 10^{-12}$	$56.6 \cdot 10^{-12}$
7.	2,3,4,7,8- chlorodibenzo-furan	NA	ND	$56.6 \cdot 10^{-12}$
8.	Dibenzo-*p*-dioxine	$1.82 \cdot 10^{-34}$	NA	-
9.	1- chrodibenzo-*p*-dioxine	$1.78 \cdot 10^{-38}$	NA	-
10.	2,3,7,8- chlorodibenzo-*p*-dioxine	$4.90 \cdot 10^{-53}$	ND	$2.3 \cdot 10^{-12}$
11.	Heptachlorodibenzo-*p*-dioxine	NA	$2.57 \cdot 10^{-12}$	-
12.	Octochlorodibenzo-*p*-dioxine	$4.18 \cdot 10^{-73}$	$5.43 \cdot 10^{-12}$	$353.3 \cdot 10^{-12}$
13.	Naphthalene	$4.00 \cdot 10^{-18}$	NA	-
14.	Chloromethane, CH_3Cl	$3.26 \cdot 10^{-6}$	NA	-
15.	Cyanic acid, HCN	$1.58 \cdot 10^{-5}$	NA	-

NA – not analyzed; ND –not detected.

An experiment on solid propellant charge combustion was made. A solid propellant charge with mass of 3.38 kg was taken, in contrast to [9], where a charge with mass 10 g was used. Estimated combustion time was 20 s, and a cooling water discharge of 3 kg/s was supplied over 25 s. Throughout the process, the temperature in the reservoir (10, Fig. 2) was not higher than 90°C.

Maximal values of mass fraction of toxic substances in solid propellant combustion products after calculation and experiments are given in table 1.

3. CONCLUSION

The research finds that super-toxic substances in toxicity equivalence TEQ 0.116 ng/kg of solid propellant. Such quantity of dioxins compares with soil background pollution in cities of Russia (international standard – 10 ng/kg of the substance). Thus, it has been proved that during solid propellant combustion dioxins and similar substances are not formed if combustion products are quickly cooled with water.

References

[1] Fiodorov, L.A. (1993) Dioxine as Ecological Danger: Retrospective View and Prospects. Nauka, Moscow.

[2] Tan, P., Hurtado, I., Neuschuetz, D. (2001) Thermodynamic Modeling of PCDD/Fs Formation in Thermal Processes //Environmental Science & Technology, 35, pp 1867-1874.

[3] Ishizu, J., Yoshihara, Yo., Hiraoka, M., Endo, K. Investigation of Dioxin Formation in Municipal Solid waste Incineration Based on Chemical Equilibrium.

[4] Kolesov, V.P., Dorofeeva, O.V., Iorish, V.S., Papina, T.S., Lukyanova, V.A., Melkhanova, S.V. (1999) Experimental Measurements and a Group Additivity Approach for Estimating the Standard Molar Enthalpies of Formation of Dioxins. Mendeleyev Communications Electronic Version, Issue 4, pp 129-170.

[5] Dorofeieva, O.V., Gurvitch, L.V. (1996) Thermodynamic Characteristics of Polychlorinated Dibensoyl-*n*-dioxine and Dibensoyl-furan in Gas Phase. Journal of Physical Chemistry. Volume 70, No. 1, pp 7-12.

[6] Ridd, R., Prausnitz, J., Sherwood, T. (1982) Characteristics of Gases and Liquids. Reference book. Khimija, L.

[7] Marinov, N.M., Pitz, W.J., Westbrook, C.K., Castaldi, M.J., Senkan, S.M. (1996). Modeling of Aromatic and Polycyclic Aromatic Hydrocarbon Formation in Premixed Methane and Ethane flames. Combust. Sci. and Tech., Vols.116-117, pp 211-287.

[8] Thermodynamic Characteristics of Individual Substances: Reference publication in 4 volumes. (1982). Nauka, M.

[9] Samsonov, D.P., Kiriukhin, V.P., Zhiriukhina, N.P., Pervunina, R.I. (1996). Determination of the Content of Polychlorinated Dibenzo-n-dioxines, Dibenzofurans, Bi-phenyls and Polyaromatic Compounds in Combustion Products of Solid Rocket Fuel. Journal of Analytical Chemistry. Volume 5, No. 11, pp1218-1221.

RISK ASSESSMENT OF CHEMICAL WEAPONS INFLUENCE ON ECOSYSTEMS AS THE WHOLE

V. N. Lystsov[1], N. V. Murzin[1], A. A. Bykov[2]
[1]RRC «Kurchatov Institute», Moscow, Russia, [2]SRC «Ecosafety», Moscow, Russia

Abstract: Methodology for risk assessment has been developed on the basis of definition of the detriment for ecosystem as the whole. Species-edificators are considered as the basic component influenced by toxic chemicals. The system of safety criteria for natural communities is proposed, where the quantitative risk scale and actual changes under influence of poisonous and toxic chemicals in community are compared.

Key words: ecological risk, natural communities, ecosystem, poisonous and toxic chemicals, chemical weapons

1. INTRODUCTION

An ecosystem as a whole is characterized by its structure, composition and energy per unit time spent for its own needs. Under the influence of poisonous and toxic chemicals an ecosystem incurs detriment, which could be expressed in natural indicators such as decrease in production of living matter and decrease in the flow of energy through the ecosystem. It is necessary to distinguish between detriment to ecosystem and detriment to its populations.

Detriment for consumer's populations is unobservable for ecosystem even in the case of complete loss of some consumers at high trophic levels, while corresponding mass and energy flow is not higher than several percents from total.

Detriment for particular population of producers can not be connected with the risk for ecosystem as the whole, while weakness of one such population will give competitive advantage to another one, which was more adaptable to new conditions. At the same time changes in composition and structure of a community take place and change of dominant species proceeds. Correspondingly detriment to edificators of ecosystem is a suitable characteristic for ecosystem as a whole.

A decrease in a food resource decreases its availability and sharpens the inter- and-intra species competition for this resource. Even an insignificant

V.M. Kolodkin and W. Ruck (eds.),
Ecological Risks Associated with the Destruction of Chemical Weapons, 65–74.

decrease in production of living biomass by edificators could be harmful for populations of high trophic levels. At the same time, direct influence of poisonous and toxic chemicals could be insignificant.

The assessment of the natural detriment to an ecosystem coming from poisonous and toxic chemicals in natural media can not be completed without consideration of the ecosystem as the whole and on the other hand without taking into account its structure and composition. Such assessment of natural detriment and, based on it, risk assessment from harmful influences, should take into account:

- Decrease in the level of life activity of organisms forming the ecosystem;
- Change in composition of the ecosystem due to loss and migration of living organisms;
- Change in structure of the ecosystem due to decrease in biodiversity and due to succession from one ecosystem to another;
- Loss of the ecosystem.

"An ecosystem is any community of living organisms and its environment unified into a functional whole, which rises as the result of interdependence and causal linkages existing between separate ecological components"[1]. Further, by "ecosystem" we mean "mesoecosystem" or biocenosis of forest, meadow, water reservoir – in other words the biocenosis occupying territory in which natural conditions are homogeneous. Such territory should have sufficient area to provide for sustainable life of all organisms belonging to such community.

"Ecosystem" defined in this way is the totality communities of smaller by size (biocenoses), which will be named here as "elementary communities" of living organisms. For most parts of mesoecosystems there exists a characteristic set of "elementary communities," similar by their structure and composition. Each "elementary community" has its own territory. These territories do not overlap and the total ecosystem territory is the sum of the territories of all "elementary communities." As an example of "elementary community" one may use a separate tree together with all accompanying living organisms – other producers, consumers and reducers, living by its resources.

The functioning of an ecosystem as a whole is the result of simultaneous functioning of all its separate "elementary communities." The place for each of them in the functioning of ecosystem is defined by biomass and energy produced and their position in the row of succession.

2. NATURAL DETRIMENT FOR ECOSYSTEM

Response of an ecosystem to external influence is characterized by the level of natural detriment resulting from such influence. Natural detriment can be defined as the decrease of biomass production in the ecosystem at any partcular time due to:

- slowing down or complete cessation of growth and development of living organisms,
- shrinking of the territory occupied by the ecosystem, and
- premature death of some living organisms in the ecosystem.

Production of matter and energy in the ecosystem at any moment in time could be characterized as the difference between full (brutto) production of biomass at the same moment and expenditures for competition, breathing and support of living activities, as well as expenditures for dying out of living matter. Production of biomass and energy accumulation is proceeding at the "basic" productive trophic level of ecosystem, which is usually formed by the primary producers. Life activity at other trophic levels is fully dependent on the level of production at the basic trophic level.

Biomass produced is partly used by consumers and partly increases the biomass of the productive trophic level in the productive type of ecosystem. In the protective type of ecosystem (climax and near climax communities in a row of succession) practically all biomass produced is used by consumers or is utilized by reducers, or can be spent for living processes.

In accordance with Le Chatelier principle [2], any living organism is able to compensate a harmful influence applied to him, if the necessary expenditures do not overcome the energy resources available to this organism. These resources are formed from:

- energy accumulated in earlier produced biomass and
- newly synthesized biomass.
- thus the compensation of the harmful effect will bring a decrease:
- in biomass production and
- stores of energy due to loss of biomass in the system.

Therefore an external influence will lower production of living matter in an ecosystem due to an increase of energy expenditure for living processes. It will also decrease biomass accumulation and a fraction of it available for consumers, but it will increase the level of spent living matter (litter) (Fig. 1).

Fig. 1 Dependence of production level on time in presence and absence of toxic pollution

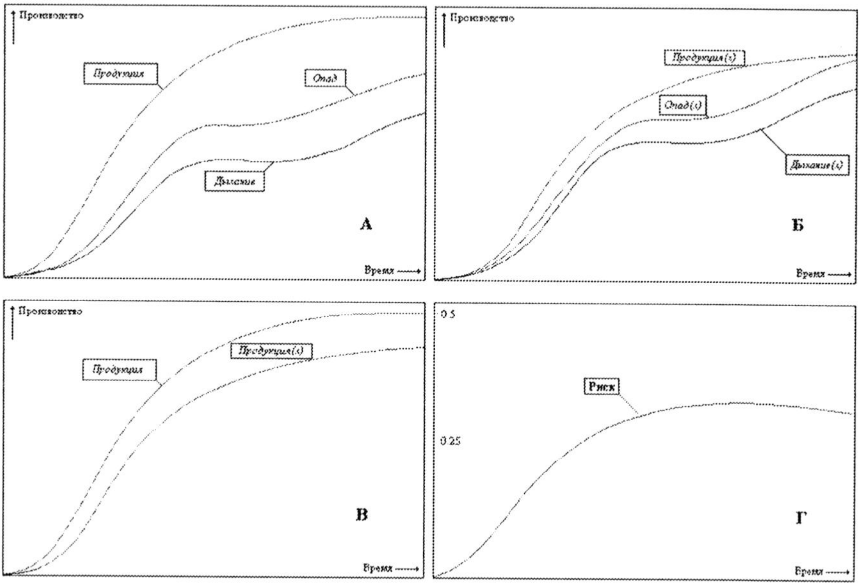

In unpolluted territory production of the community is equal to the difference between total production and breathing expenditures (A). In the presence of pollutants, the total production is decreased and breathing expenditure increases (Б). The quantity of litters also increases. Net production available for producers and consumers diminishes from B to $B(x)$. The natural detriment to the community, as a deficit of biomass per unit area of biocenosis, could be defined as the difference between these two values (B), and risk as the ratio of this difference to the maximum level of production (Г).

With increasing influence of pollutant, the fraction of resources available to consumers will shrink and it will cause first consumers of high trophic levels, and later phytophages, to drop out of the ecosystem. Resources available for reducers will increase in absolute value but in terms of fraction of available resources such increase will be insignificant. Possibly the increase of this fraction will be sufficient for compensation for the increased harmful influence.

Change in the state of "elementary community" dB will depend on the difference between energy utilized by community dE and value of energy dQ, which was directly spent for support of the necessary level of living activity of community during time dt:

$$\frac{dB}{dt} = A \cdot \left(\frac{dE}{dt} - \frac{dQ}{dt} \right), \qquad (1)$$

where A is a transfer coefficient showing which fraction of energy will be transformed into production of "elementary community."

In the difference from standard ecological characteristics of ecosystem, the value B characterizes the "instant" state of the community at any moment in time and takes into account losses for competition and support of the sustainable state of the community. If standard ecological characteristics describe the state of ecosystem at the end of vegetation season, the definition of function B permits following the dynamic changes of the state of the "elementary community" during vegetation season.

Change in the state of "elementary community" dB in time moment t during the vegetation period can be calculated as the difference between energy utilized by "elementary community" and energy spent at all its trophic levels. If one supposes that the first of fermentative reactions in the system of biomass synthesis and the first of the dissimilation reactions are described by Michaelis-Menten cinetics [5], then the equation for rate of change in the state of the "elementary community" could be presented in the following form [2,4,6,8]:

$$\frac{dB}{dt} = V_M(x_n) \cdot B \cdot \left\{ \frac{y}{1+y} - \frac{d(x_n) \cdot y}{1 + 2 \cdot L + y} - \frac{B}{B_{M0}} \right\}, \qquad (2)$$

where:

x_n level of influence on the "elementary community" n,

y dimensionless concentration of limiting biogen, expressed as the fraction of low limit of tolerance zone for this biogen,

L dimensionless half-width of tolerance zone, measured as the difference between high and low limits of tolerance zone,

$V_M(x_n)$ dimension of the "elementary community" at the level of influence x, day^{-1}, which will be considered below,

$d(x_n)$ loss factor for living activity at all trophic levels of the "elementary community",

B_{M0} maximum level of $B(t)$ during the season for the "elementary community." B_{M0} characterizes in particular the losses from competition.

In the presence of a toxic influence on the "elementary community" (x_n>0) its state could change due to:

- inhibition of fermentative reactions by the toxic chemical and/or decrease in fermentative activity, which will appear as decrease in the specific rate of production;
- increase in mortality as well as breathing losses in living organisms in "elementary community", which will appear as increase in loss factor $d(x_n)$.

As the result of all these processes there will be a decrease in the available volume of habitat, which will be expressed as decrease in the possible maximum level of production in the "elementary community."

The action of toxic chemicals influences the metabolism of living organisms by decreasing the specific rate of production $V_M(x_n)$ down to zero at high levels of exposure and generally increasing in loss factor $d(x_n)$ up to limit value d_M.

Keeping in mind that the first of these parameters is connected with biomass production in the system of metabolic reactions, and the second one with dissimilation of produced matter, one can characterize the influence of exposure to x_n on the life activity of living organisms as inhibition or activation of corresponding fermentative reactions:

$$V_M(x) = \frac{V_M(0)}{K_V + x_n} \cdot K_V, \tag{3}$$

$$d(x_n) = \frac{d_0 \cdot K_D + d_M \cdot x_n}{K_D + x_n} = d_0 + (d_M - d_0) \cdot \frac{x_n}{K_D + x_n} \tag{4}$$

where

K_V constant of inhibition of biosynthesis processes,
K_D activation constant of processes of breathing and dissimilation,
d_0, d_M value of loss factor in the absence of influence, and the maximum value of loss factor,
$V_M(0)$ maximum specific rate of production, calculated with the use of ecologo-physiological parameters of organisms (see, e.g. [2, 3, 4, 9]) or experimentally defined (see [7]).

Data on the values of inhibition and activation constants could be obtained on the basis of analysis of the influence of poisonous and toxic chemicals on living organisms, which may be characterized by 50% lethal values LC_{50} and LD_{50} [9].

Natural detriment $H_n(x_n,t)$ to the "elementary community" n, expressed as the dry biomass not obtained by the community from the unit area in time moment t during the vegetation season at the level of influence x_n is equal to the difference between the state $B(0,t)$ of the "elementary community" in the absence and in the presence of influence $B(x_n,t)$ for $x \neq 0$:

$$H_n(x_n,t) = B(0,t) - B(x_n,t). \qquad (5)$$

While different "elementary communities" are differentiated by their characteristics, natural detriment, expressed as above, is inconvenient for analysis and normalization. Instead it would be better to consider relative detriment $R_n(x_n,t)$ comparing detriment in relation to maximum level of production $B_M(0)$, and this expression may be considered as the "danger index" (risk) for the "elementary community":

$$R_n(x_n,t) = \frac{B(0,t) - B(x_n,t)}{B_M(0)}. \qquad (6)$$

For calculation of this danger index under the influence of poisonous and toxic chemicals one needs to know the whole set of parameters used in the equations. It is convenient to discern two groups of parameters:

- Ecophysiological
- Ecotoxicological.

All these parameters could be directly measured or calculated from measured value. For most parts of the "elementary communities", however, the calculation of the index of danger could be made on the basis of parameters of a hypothetical "typical community" and information on the current level of pollution.

3. RISK ASSESSMENT FOR ECOSYSTEMS

Risk is usually understood as the probability of negative consequences and the size of these consequences. Very often mathematical expectation is used as the quantitative measurement of risk, or in other words it will be the expected detriment. Risk to the ecosystem may be defined just in this way, as the expected detriment to the ecosystem on the territory under external influence in comparison to maximum level of production of the same ecosystem on unpolluted territory.

Risk *R(x,t)* for the ecosystem as the whole at the level of exposure of the "elementary community" x_n at any moment in time t during the vegetation period is defined by the expression:

$$R\ (x,t) = \frac{\sum_{1}^{N} w_n \cdot R_n(x_n,t)}{\sum_{1}^{n} w_n}, \qquad (7)$$

where

$R_n(x_n,t)$ danger index or risk for the "elementary community" n in time moment t,

w_n weight coefficient for the "elementary community" n in ecosystem, equal to relative significance of the "elementary community" n in ecosystem composed out of N "elementary communities" of different times, $0 \leq w_n(t) \leq .1$.

Thus, risk for an ecosystem under direct influence of poisonous or toxic chemicals on living conditions of the "elementary communities" in time moment t can be defined as the weighted average, taking into account relative significance $w_n(t)$ of all "elementary communities" and danger index, of all these "elementary communities".

If the ecosystem is homogenous by composition of its "elementary communities", then contributions of each community in any time moment will be equal and then:

$$R(x,t) = \frac{1}{N} \cdot \sum_{n} R_n(x_n,t). \qquad (8)$$

The consequences for consumers, after impact of the influence on the ecosystem, should be defined by taking into account the risk for all "elementary communities" localized on the territory occupied by the population.

The relations formulated above allow the establishment of safety criteria for different ecosystems according to the level of risk (Table 1). The safety criteria for the natural community of the forest succession of the temperate zone, collected in Table 1, could become the basis for decision making on the safety from influence of poisonous chemicals on the natural communities at the sites of storage and destruction of chemical weapons.

Table 1: Approximate safety criteria for terrestrial community of forest succession of temperate zone

№	*Level of risk higher than*	*Community position in the succession row*	*Content meaning of the criterion*
1	0.8	Beginning of succession	Possible death of community, patchiness of coverage, low biodiversity, absence of successional development
2	0.7	Before climax	Possible death of community and/or change to the younger community in successional row.
3	0.6	Climax	Possible death of climax community and/or change to community before climax
4	0.1	Beginning of succession	Change of the community structure/composition, decrease in consumers biodiversity including phytophages and saprophages
5	0.02	Before climax	
6	0.01	Climax	
7	0.01	Beginning of succession	Loss of rare and/or highly specialized species of zoophages
8	0.002	Before climax	
9	0.001	Climax	

Safety criteria for terrestrial communities of steppe succession and for freshwater communities of temperate zone are presented in [8]. Approximate values of ecological parameters, which are necessary for assessment of the influence on ecosystems of concrete poisonous chemicals, are discussed in [9].

References

[1] Reimers N.F. (1990). Nature use. Handbook, Mysl, M

[2] Odum Yu. (1975). Basic ecology, Mir, M.

[3] Ricklefs R. (1979). Basis of general ecology, Mir, M.

[4] Fedorov V.D., Gilmanov T.G. (1980). Ecology, MSU, M.

[5] Dixon M., Webb E. (1982). Ferments, 3 vol., Mir, M.

[6] Bykov A.A., Murzin N.V. (1997). Problems of safety for man, society and nature, Nauka, S.-Pb.

[7] Prosser L., Editor (1977). Comparatiive physiology of animals, Mir, M.

[8] Murzin N.V., Bykov A.A. (2003). Calculation of the detriment for natural communities and risk for ecosystems from the influence of poisonous and toxic chemicals, «Voprosy analisa i upravlenija riskom», 1 (No.1): 32-50.

[9] Murzin N.V., Lystsov V.N., Bykov A.A. (2003). Assessment of the influence of poisonous and toxic chemicals on ecosystems, « Voprosy analisa i upravlenija riskom », 1 (No.1): 51-78.

HEALTH AND ENVIRONMENTAL RISKS ASSOCIATED WITH THE DESTRUCTION OF CHEMICAL WEAPONS

Jiri Matousek[1]
[1]*EU Research Centre of Excellence for Environmental Chemistry and Ecotoxicology, Faculty of Science, Masaryk University Brno, Czech Republic*

Abstract: The provisions of the Chemical Weapons Convention (CWC) connected with possible contact of personnel are summarised. Any handling with chemical weapons (CW), both in bulk and in munitions as well as with toxic armaments waste, in implementing the CWC, including verifying its provisions, is associated with actual health and environmental risks. The most fundamental provisions of the CWC, which reflect its original purpose,, are those dealing with destruction of CW, which present a high risk during the whole course of destruction operations, starting with loading the cargo containers, over transportation, filling of destruction equipment, etc. untill the disposal of non-toxic waste and scrap metal. The developed and operational destruction technologies are reviewed. The system of workplace safety and environmental protection of CW destruction / disposal operations including their verification as envisaged by the CWC is to be based on toxicological data of toxic agents to be destroyed, efficiency of technologies and toxic properties of end products. At present, with the exception of studies connected with the influence of sea-dumped CW on aquatic environment, there is a lack of data dealing with the impact of CW on other environmental components and ecosystems.

Key words: Chemical Weapons Convention, CW destruction technologies, health risks, environmental risks, safety system

1. INTRODUCTION

Any operations associated with the implementation of the Chemical Weapons Convention (CWC) and verification of its provisions is connected with health and environmental risks, as we can shortly summarise in table 1. It is obvious that the destruction and disposal of chemical weapons, starting with closing and sealing storage facilities, location and excavation of ancient buried munitions, loading and unloading cargo containers, transport, filling of destruction equipment, controlling destruction processes and disposal of end-products, as well as on-site verification of all mentioned activities is

V.M. Kolodkin and W. Ruck (eds.),
Ecological Risks Associated with the Destruction of Chemical Weapons, 75–83.

inherently connected with health and environmental risks. It therefore needs a sophisticated system of means and measures in order to assure required level of mainly indoor (workplace) safety and environmental protection. Such system should encompass safety standards such as concentration/exposure limits, emission/imission limits for toxic agents and end-products, and other relevant toxicological and ecotoxicological standards, analytical means and techniques for detection, determination of toxic agents and products of their destruction, monitoring means for alarming and early warning in case of leakage and/or incidents and accidents, instruments for controlling destruction processes, decontamination methods and means, means and methods for protection, prophylactics, first aid, therapy and medical rescue system.

Table 1: Short overview of the risk activities predicted by the CWC [1]

- closing, sealing, visiting storage sites,
- checking declarations,
- any operations and on-site inspections at destruction/conversion of former production facilities,
- any handling and on-site inspections of acting small-scale production facilities
- any handling and on-site inspections at CW destruction facilities, including all operations starting with loading cargo containers at storage sites, transport, filling of destruction equipment, controlling the destruction process, checking completeness of destruction, till the safe disposal of non-toxic waste, scrap metal etc.,
- on-site inspections on challenge,
- protection against and verification of alleged use of chemical weapons,
- any handling and on-site inspections at extremely dangerous operations connected with location, excavation, transport, demilitarisation and destruction of old and abandoned chemical weaponry and toxic armaments wastes .

2. TOXIC AGENTS TO BE DESTROYED AND PROPOSED / APPLIED TECHNIQUES

In comparison with various reports on the number of possessors of chemical weapons still from the 1980s, obviously politically influenced in the last decade of the Cold War and East-West confrontation, the good news upon entering the new Millennium is that this number does not exceed the number of fingers of one hand, at least among the Party States (PS) to the CWC according to their declaration on CW possession. The representative list of toxic agents under consideration is given by remaining stockpile in the USA and Russian Federation, the only major possessors of chemical weapons. Main US agents to be destroyed are GB, VX, HD, L, T and DM. This situation is similar in Russia, having among G-agents beside GB also huge stocks of GD, not possessing the T-agent (possessing instead a special type of H for winter conditions), still having the HL-mixture and another type of V-agent, with the same molecular mass like VX (the O-isobutyl N,N-diethyl analogue to VX, named simply V-gaz in Russian), and a small amount of already obsolete agents (like CG) [2]. Declared stockpiles by other two SPs (India and South Korea) as well as old and abandoned CW declared by respective SPs involve an assortment of the same or classical agents respectively. Research works on technologies for destruction/disposal of chemical weapons (CW) both in bulk and in munitions started long before the commencement of negotiations on the general and comprehensive ban of chemical weapons, the reason for this is that such operations are not only an imperative of the prohibition and elimination of chemical arsenals once and for all, but they were connected with the upgrading of chemical weaponry under increasing awareness of environmental protection. This new found environmental protection no longer allows such barbaric procedures for destruction/disposal, such as ocean dumping, earth burial and open-pit burning, typical still after the WW-II mainly when destroying German and Japanese CW under Allied supervision. The last known event of this type was the operation CHASE, i.e., dumping of the US nerve agent munitions in the Caribbean during the 1970s. Beside the main technological procedure (i.e. incineration), there is just a relatively long list of various technologies at different stages of realisation. The overview of suggested technologies for destruction of CW is presented in Table 2

The list of destruction techniques also encompasses some curious procedures, such as the underground nuclear explosions, certainly very effective but hardly acceptable because they pose other risks and the impossibility of checking the completion of the destruction process as required by the CWC, or the otherwise attractive proposal by N. Platé, offering the possibility to destroy two kinds of armaments simultaneously, i.e., CW and missile engines, pursuant to the bilateral US-USSR IMF agreement. It can be noted that according to the theoretical considerations or/and actual experience with operational technologies, there are already attempts to assess the available and potential alternative technologies taking into consideration all technological, economic as well as the health aspects [4,5]. These studies are aimed at proven, realistic technologies. The interesting review assessment contains the only study on alternatives to incineration technologies, dealing with toxicity of end-products [6].

Table 2: Technologies for destruction of CW stockpiles [3,4]

- Two Stage Technology (Soviet) – chemical deactivation + detoxification
- Incineration, e.g. JACADS (Johnson Atoll Chemical Agent Disposal System – USA)
- Underground nuclear explosion – Arzamas-16 according to Academician Trutnev
- Disproportional use of liquid propellant missile engine together with the elimination of missile engine according to Academician N.A.Plate
- Biological methods – Biodegradation of Organophosphates
- Supercritical Water Oxidation
- Wet Air Oxidation
- Molten Salt Processes
- Photochemical Processes:
    - gaseous phase process (photochemical)
    - aqueous phase process
    - ozone/ultraviolet irradiation
    - laser-stimulated photodegradation
    - catalytic ozone oxidation, gamma irradiation
- Electrochemical techniques
- "Neutralisation": hydrolysis, oxidation etc.
- Chemical reprocessing: chlorolysis, catalytic dehydrochlorination
- Thermic processes:
    - plasma reactors
    - microwave plasma process
    - infra-red thermal process
    - radiofrequency thermal process
- Solvated electron technology

3. HEALTH RISKS OF CW DESTRUCTION TECHNOLOGIES

According to the profound toxicological studies (obviously only by extrapolation from data accumulated in experimental animals), the maximum allowable workplace and outdoor concentrations for principal CW agents have been suggested in both US and Russia. The values, shown in Table 3 are based on the LOAEL and NOAEL approach and inhalation of contaminated air. The problem is, how to detect and monitor the presence of toxic agents in the workplace and ambient atmosphere. Comparing the indoor limits with the detection limits of existing devices of chemical reconnaissance, based on detection tubes, indicator papers and simple field laboratories and detection kits could lead to the conclusion that such equipment items are useless for this purpose. Nevertheless, our critical assessment [3] shares the view of e.g. Chebotarev et al. [7] that such equipment is not only available everywhere and represents a suitable base for monitoring at the storage and destruction site. The dangerous handling must be assured by technical measures at the side of the technological equipment by its hermetising and fitting with corresponding barriers and containments, as well as on the side of the crew by its physical protection, and last but not least, by the telemetric control system etc. It means that the cases of use of standard military devices, not only mentioned above but also of automatic monitors, which are principally designated to detect battlefield concentrations are incompatible with those considered as hygienic standards for workplace.

Table 3: Russian and US standards for maximum allowable concentration of the main CW agents (mg/m^3) in air [3.4]

	Russia		**USA**	
Chemical agent	Work area limit	GPL[a]	Work area limit	GPL[a]
Sarin (GB)	$2x10^{-5}$	$2x10^{-7}$	$1x10^{-4}$	$3 x10^{-6}$
Soman (GD)	$1x 10^{-5}$	$1x 10^{-7}$	$2 x10^{-5}$	$3 x10^{-6}$
VX	$5 x10^{-6}$	$5 x10^{-8}$	$1 x10^{-5}$	$3 x10^{-6}$
Mustard (HD)	$2 x10^{-4}$	$2 x10^{-6}$	$3 x10^{-3}$	$1 x10^{-4}$
Lewisite (L)	$2 x10^{-4}$	$4 x10^{-6}$	$3 x10^{-3}$	$3 x10^{-3}$

[a] General Population Limit

Safety protection only breaks down when an accident occurs. In contrast, air monitoring in the personnel protection zone, which is a zone affording a manifold protection, is made over a longer period of time, which allows long-term sampling of the air, in order to check if the indicated limits are reached and if the upper limit turns out to be an average concentration.. For the outdoor limits, the concentrations are generally lower by 1 to 2 orders of magnitude than the limits shown for indoor (workplace) environment. This means that they would still be much more difficult to detect by means of any regular chemical reconnaissance device or equipment item. On the other hand, it is obvious that such standards are valid for long-term repeated exposure that would be the case if a continuous toxic plume was being released as a result of the destruction of a facility or storage site.Yet such facilities are principally constructed so as to render such an incident highly improbable. Nevertheless, the reason for installing monitoring devices based on enzymatic inhibition or on any other physicochemical principle, such as the ionization principle or ion mobility, or nowadays even the most sophisticated monitors based on long path or remote (off-site) sensing, is to watch both indoor and outdoor atmosphere in the case of incidents or accidents to make sure indoor personnel are warned and outdoor population so that they can adopt previously practiced and planned emergency measures. [3] Any of the verification exercises anticipated by the Chemical Weapons Convention is connected with the utilisation of objective analytical methods involving air monitoring. It is possible to say that the verification carried out as on-site inspection at facilities for CW destruction, including all operations beginning at storage sites, i.e., safe loading to cargo containers, transport to the destruction facility, filling into the destruction equipment and verifying completeness of the destruction process, generally do not count among physically easy tasks. On the other hand, from the analytical point of view, they are by far not as complicated as the tasks last indicated in table 2. It is generally not necessary to carry out something like systematic analysis of unknown sample, the known, i.e., declared, compounds can be detected in large amounts, so that for these purposes, even very simple detection equipment can be used. This procedure, however, is connected with high dangers for inspectors due to (expected) leakage of unitary munitions. Therefore, air monitoring within the framework of the verification will be at the same time a measure of workplace safety from the point of view of inspector, and of course, of handling personnel. Present military monitors, alarms, analysers and other devices for chemical reconnaissance, possessing any physical, physicochemical, biochemical and biological principles and most probably even instrumental techniques of foreseeable future, assuring on-site and remote sensing, cannot reach the stringent maximum indoor and especially outdoor permissible concentrations to give immediate response in real time on the actual ambient concentrations of the super-toxic lethal agents in the concentration range, corresponding above mentioned concentrations. Present relevant analytical technology, as mentioned above, as well as such technology as anticipated for the time space of CW destruction according to the Chemical Weapons Convention is, however, able to assure incidental (accidental) early warning against lower than threshold concentrations, for periods of time (exposure) of hour(s) or higher, allowing to accept necessary emergency measures (including protection) in necessary time frame. Besides detection and monitoring of toxic agents' presence, means for personal protection, decontamination first aid and qualified medical aid also belong to the system.

4. ENVIRONMENTAL PROBLEMS OF CW DESTRUCTION TECHNOLOGIES

All hygienic and toxicological standards dealing with actual CW agents, as mentioned above, have been derived from experimental toxicology because there are not enough data from accidental intoxication of humans comparable with those stemming from epidemiological studies on much less toxic chemicals. Similarly, with the exception of the frequent toxicological data on experimental animals (suggesting for very high acute toxicity), generally, and for obvious reasons, not enough relevant ecotoxicological data are available, which include information on the impact of CW agents on ecosystems. There is perhaps only one exception, that is the studies devoted to the fate of sea-dumped CW inventories of the Nazi and Japanese imperial armed forces after WW-II and studies on the behaviour of CW agents in sea water, repeatedly published in the 1960s and 1970s in the connection with the last disposal operations of this kind. This is the reason why it is possible to find ecotoxicological studies in aquatic organisms in connection with the environmental impact of sea-dumped CW inventories, mainly in the Baltic and adjacent areas, such as Skagerrak and the Norvegian trench [10, 11, 12].

5. CONCLUSIONS

All operations connected with the destruction and its verification, starting with any handling on storage sites, ranging from handling with occasionally found and improperly disposed-of old and abandoned munitions, over transportation and proper destruction to the disposal of non-toxic waste and scrap metal, pose health and environmental risks. This is rather due mainly to leakages and releases at accidental events (including sabotage and terrorism) than at own destruction operation at CW destruction facilities where an appropriate system of workplace and outdoor health safety measures is to be prepared just in their design. Environmental risk mainly depends on the toxicity of end-products and their disposal.

References

[1] Matousek J.(1994). Workplace safety at the operations connected with the implementation of the Chemical Weapons Convention. Second Moscow International Conference on Chemical Disarmament, MOSCON 94, Moscow.

[2] Beletskaya I.P. and Novikov S.S. (1995). Russian chemical weapons. Herald of the Russian Academy of Sciences 65: No 1, 5-10.

[3] Matousek J.(1997). Methods and means for air monitoring associated with the destruction of chemical weapons. In: M.Heyl and R.McGuire (Eds.), Analytical Chemistry Associated with the Destruction of Chemical Weapons. Kluwer Academic Publishers, Dordrecht, pp. 181-187.

[4] Hart J., Miller C. (Eds.) (1998). Chemical Weapon Destruction in Russia: Political, Legal, and Technical Aspects. Oxford University Press, Oxford.

[5] Pearson G.S., Magee R.S.(2002). Critical evaluation of proven chemical weapon destruction technologies. Pure Appl.Chem. 74: No 2, 187-316.

[6] Picardi A., Johnston P., Stringer R. (1991) Alternative Technologies for the Detoxification of Chemical Weapons. Greenpeace International, Washington.

[7] Chebotarev O.V., Druzhinin A.A., Pashinin V.A. and Sinicyn A.N. (1994). Express-analysis on the objects of storage and destruction of CW utilising regular means of chemical reconnaissance and chemical control (in Russian). Russian Chemical Journal 38: No 2, 69-73.

[8] Stock T., Lohs K.(Eds.) (1997). The Challenge of Old Munitions and Toxic Armaments Wastes. Oxford University Press, Oxford

[9] NATO/CCMS (1995). Cross-Border Environmental Problems Emanating from Defence-Related Installations and Activities, Vol.2. Chemical Contamination. Report No 205. NATO, Brussels.

[10] Gorlov V.G. et al. (1993). Complex analysis of the hazard related to the captured German chemical weapons dumped in the Baltic Sea. National Report of the Russian Federation, Moscow.

[11] Wibberenz G.(1992). Gefährdungen durch Giftgas in der Ostsee. PFK, Kiel.

[12] Matousek J.(2002). Old scrap munitions and toxic armaments wastes in the Central European and Baltic Region. SECOTOX – 7th Regional Meeting of the Central and

East European Section:. Trends and Advances in Environmental Chemistry and Toxicology , Brno, Proceedings pp. 246-252.

ECOLOGICAL RISKS ANALYSIS FOR THE CHEMICAL WEAPONS DESTRUCTION FACILITY

T. Shvetsova-Shilovskaya[1]

[1]*State Research Institute of Organic Chemistry and Technology (SRIOCT), Moscow (Russia)*

Abstract: The problem of risk analysis and assessment in emergency situations at hazardous chemical processing facilities has become especially acute in the recent past. The European document providing for control over safety of chemical processing facilities is the Sevezo Directive. It has become the basis for modern legislation in European governments on industrial and transportation safety. Over the past few years, Russia too has intensified its development of a legislative and normative base in the sphere of industrial safety. The "Act on the Industrial Safety of Hazardous Production Facilities" was adopted (1997). In accordance with this Act, each hazardous facility in Russian Federation is to develop a Safety Declaration that is subject to expert review, and obtain a production permit from respective authorities. This article presents the Russian approach to the analysis of the extra hazardous chemical processing facilities, namely the chemical weapons destruction facilities (CWDF). It is shown that declaration is an integral part of ensuring industrial safety and necessary for the estimation ecological danger posed by CWDF and for the purposes of ecological insurance.

Key words: safety declaration, risk analysis, chemical weapons destruction facility, ecological insurance, reliability of equipment.

1. INTRODUCTION

The experience of the European countries shows that the declaration of safety of industrial facilities allows lowering, essentially, the number and harm of possible consequences of failures and extreme situations generated by them. According to the «Act on the Industrial Safety of Hazardous Production Facilities» of the Russian Federation (June 20, 1997), declaration is an integral part of maintenance of industrial safety and the prevention of failures at dangerous industrial facilities.

It is necessary to emphasize, that this Act is based on the requirements of the «Sevezo Directive». One of its main positions is the preparation of safety declarations. The structure of the safety declaration includes the following:

V.M. Kolodkin and W. Ruck (eds.),
Ecological Risks Associated with the Destruction of Chemical Weapons, 85–89.

1. General data about an industrial facility. 2. Results of the safety analysis. 3. Provision of requirements of industrial safety. 4. Conclusions.

The first section comprises brief data on the industrial facility, namely full and the abbreviated name of the facility, the address, the name of the head, the description of the facility location, purpose of the facility, dates of construction and reconstruction, data about the developer and the builder. The description of the facility location contains the name of the region, settlement or distances to it, names of nearby reservoirs, the size of a sanitary and protective zone, the topographical data, data on climatic conditions; the data on the personnel working on the facility and population living close. There is some data on recipients which can suffer from the results of an accident due to a failure in the declared facility. In the given subsection attributes of the contract of insurance are also specified.

The second section contains the information on technology and hardware registration, on characteristics of dangerous substances. The basic results of the Risk Analysis are stated. The data are given on the sizes of zones of dangerous factors, on possible number of victims, on possible damage to the population, property and the environment.

The third section reflects features of the organization of external and internal supervision of safety. Here the description of structure and duty of the service supervising industrial safety, a quality monitoring and supervision, the list of documents on which the control and supervision, structure and functions of counter-failure forces and rescue services are given as well as the data on equipment of rescue services and counter-failure forces. This data includes the description of technical and organizational decisions and the actions aimed at safety. The description of the system of notification in extreme situations is given. Available local systems of notification conform to the requirements of the Russian Federation Government.

The fourth section includes the generalized estimation of the level of safety with the indication of the most dangerous components of the declared facility and the most significant factors influencing risk parameters.

2. RISK ANALYSIS

The main part of the safety declaration is the subsection with the calculation, i.e., the explanatory note to the declaration of safety, «The Risk Analysis». It includes data on known failures, the analysis of conditions of occurrence and the development of failures, definition of typical scenarios of occurrence and development of failures, a substantiation of physical and mathematical models and the methods of calculation used for the estimation of risk, calculation of probable active areas of dangerous factors, an estimation of the possible number of victims, an estimation of possible damage, and an estimation of risk of failure. Risk is a many-sided concept. Risks generated by failures at businesses can be distinguished by the form of dangers, character of sources of risk, recipients of risk, scales of a zone of mitigation, and units of measurements. According to the kinds of risk there are also measures of risk. The most widespread one is an emergency risk for one person – individual risk and social risk. All kinds of risk can be represented by numbers (dot estimations) or graphically. Results of the analysis of individual risk are represented on a cartographical basis as lines of equal values of risk. The analysis of individual risk does not allow judgment of scales of possible failures in facilities of destruction of chemical weapons. This parameter is frequently used as a measure of potential danger for industrial targets. It is important to know this parameter for solving problems posed by zoning of territory around dangerous facilities. The results of social risk are represented as charts, so-called F/N–diagrams. This parameter characterizes scales of possible emergencies and consequently it should be used while preparing the safety statements.

The Declaration section of note contains quantitative estimation of potential danger from chemical weapon destruction facilities. Thus the following tasks are solved: 1) all possible scenarios of occurrence and development of a failure are compiled; 2) the estimation of frequency (probability) of occurrence of each scenario is given; 3) fields of source terms for various scenarios of failure development are compiled; 4) consequences of effect on a person and/or material objects, from source terms of failures, are estimated. For each of the listed problems its own methods are used.

The reasons for failures in industrial facilities can be: 1) malfunction of equipment; 2) breach of containment; 3) personnel mistakes; 4) external emergency events. For the listed reasons there are methods, which allow determination of the frequency of occurrence of failures.

As a rule, the analysis of possible malfunction of the equipment is carried out with the help of fault trees. The development of fault trees is based on full understanding of operation of considered stages of technological process of destruction of chemical weapons, kinds of possible malfunction of elements and/or systems of these stages. The fault tree graphically represents logic connections between malfunctions of the equipment and emergencies. The opportunity for carrying out the qualitative or quantitative analysis of reliability of a facility is provided. Taking into account the complexity of the equipment used in CDWF, the estimation of reliability of separate systems, and the search for ways to increase reliability and productivity both at designing, and at operation, of facilities are the priorities.

Creation of unique disarmament equipment demands application of such methods of analysis and calculation of reliability, which would allow objective accounting for features of operation, to predict reliability as early as the design stage of a facility, to exclude an opportunity of a catastrophic outcome of emergencies for people and the environment.

The analysis of CDWF failure consequences to the personnel, the population and the environment assumes construction of fields of source terms. It is a complex and rather difficult problem. However only the quantitative estimation of danger allows obtaining of an objective picture of kinds of risk of the facility.

In SRIOCT the methods and the software for the analysis of danger and an estimation of risk at facilities for destruction of chemical weapons are developed. The developed techniques and programs do not contradict the techniques recommended by federal bodies of Russia. For each scenario of possible failures parameters of damage are examined. As parameters of damage the following items are considered: 1) various kinds of damage to the personnel of the facility and the population; 2) damages to the equipment; 3) damages to the environment. Calculation and analysis of the resulting parameters of risk allow quantitative description of the level of safety of the facility. The results are the proof of observance of norms and requirements of the national legislation. SRIOCT is the pioneer in the field of development of safety declarations of facilities for destruction of chemical weapons. At present several declarations for different facilities (Shchuch'ye, Kambarka, Kizner) have been prepared.

3. APPLIED PROBLEMS

Preparation of safety declarations can be used for the solution of some applied problems: development of CDWF certification, development of the program of reduction of off-schedule losses and programs of insurance protection. The certification is necessary for creation of the description of a condition of facility coordinated with the enterprise before a failure.

Damages due to failures are off-schedule losses, which cannot be stipulated in the budget of the enterprise because of uncertainty of time of occurrence of the failure and the possible consequences. For calculation of the sizes of the loss in case of failures and industrial malfunctions the business uses the data of accounts department. The purpose of any business after a failure is the restoration of the technological process; therefore a calculation of losses incurred to achieve restoration cost is more appropriate. The cost of an industrial facility restoring property significantly surpasses the balance. Declaration of safety allows determination of typical failures and off-schedule losses, development of the list of possible preventative actions, and calculation of expenses for their occurrence. It will allow determination of a level of reasonable investments. The experience of foreign researchers indicates that it is necessary to develop a program of reduction of off-schedule losses for each business. These actions can be financed both from the budget of the enterprise and from reserves of the insurance company.

Insurance cover is a financial tool, which allows the minimization of the financial risk by transferring non-scheduled expenses for indemnity for emergency losses into the category of planned and acceptable expenses through insurance payments. From the economic point of view, risk is an expected level of losses resulting from an extreme situation. According to the law, an example of mandatory insurance is the liability insurance covering damages caused by the operation of a dangerous industrial facility. However, the questions of maximum degree of responsibility and the adequate amount of the insurance premium remain unanswered. These problems can be solved by using the results of the Risk Analysis carried out while developing the safety declaration. The results of the Risk Analysis form the basis for competent dialogue between the insurance and the insurer. Both are aware of the dangers faced by insured facilities. The efficiency of the insurance cover is determined by the quality of preparation of the system of the initial data on kinds and sources of dangers, methods of analysis, and estimation and management of risks.

4. CONCLUSION

The process of deciding on the acceptability of a CWDF from a safety point of view is of vital importance at the design, construction and operation stages. This decision is expressed in the form of the safety declaration that includes a thorough analysis of the operation of the facility. This quantitative assessment of the CWDF risk level is an objective indicator of possible accident hazards from the facility. Having such an assessment at hand provides a tangible proof of compliance with the provisions of both national legislation and international practice for the development of a program to increase industrial safety and ecological insurance.

AN APPROACH TO ASSESSMENT OF CHEMICAL RISK ON THE BASIS OF GENERATION OF TIME SERIES

Aleksey Murin
Research Institute of Natural and Technogenic Disasters, Udmurtia State University, Izhevsk (Russia)

Abstract: The work considers a formal model for assessment of the systematic and accidental risks associated with chemically dangerous compounds. It is shown that, when analyzing risk, it is necessary to take into account, in one form or another, the unsteady nature of physical-chemical processes and the negative effects due to them.

Key words: model, risk, accident, unsteady process

1. INTRODUCTION

In recent years the problem of techogenic and chemical safety, in particular, has become increasingly topical. The matter is that, on the one hand, mankind realized the threat of the effects due to natural and anthropogenous disastrous events. On the other hand, the development of the fundamental sciences reached the level necessary for a substantial analysis of this field. It is possible to speak about the birth of a new exact science: mathematical theory of safety and risk. Though a number of scientific publications appeared recently, in the author's opinion, sufficient clarity has not been achieved so far in many questions.

One of the interesting questions requiring profound study is the question of the unsteady character of risk. The point is that the physical-chemical processes accompanying disastrous processes and responsible for the negative effects are unsteady by their nature. Therefore, the unsteady character should reveal itself in the risk values describing the quantitative parameters of various types of danger.

The goal of this work is to make an attempt to build a formal mathematical model of risk as applied to the problem of environmental safety associated with the recycling of chemically dangerous compounds. The work develops the author's view on the physical fundamentals and mathematical theory of chemical risk.

V.M. Kolodkin and W. Ruck (eds.),
Ecological Risks Associated with the Destruction of Chemical Weapons, 91–98.

2. STATEMENT OF PROBLEM

Considered are the system $\Sigma = O \bigcup E$ comprising a chemically dangerous object O and a surrounding ecological system $E = M \bigcup G$ where M is the environment transmitting accidental effects, G are certain recipients of the effects. The object O influences the recipients G through the environment M.

Let the state O be described by a set of parameters $p=\{p_i\}$, $p \in \mathbf{P}$. For example, the flows of mass of specific chemical agents and compounds and also the flows of energy in various forms generated by O can be specific parameters p_i. The state of O at various moments in time $t \in [0,T]$ will be characterized by various values of the parameters, i.e., $p=p(t)$ is a certain process.

Then, let the state of the transmitting environment M be characterized by the set of parameters $m=\{m_i\}$, $m \in \mathbf{M}$. For example, the wind speed and direction, temperature, class of the atmospheric stability and others will be such parameters in the particular case of transmission of the chemical effect by means of the free air. The state of M in various time moments will be characterized by various parameters, i.e., $m = m(t)$ is a process.

Finally, let the state of recipients G be characterized by the scalar parameter ξ [illegible] which is a certain function of action (losses or damage). Assume that the parameter ξ is entirely determined by the state of the sub-system $O \bigcup M$

$$\xi = \xi(p(t), m(t)). \tag{1}$$

For example, consider a case of toxic effect. Let ξ be a probability of injury to an individual at a space point x. The probability of injury due to the inhalation effect of the concentration field $c(x, t)$ depends on the toxic load $D(x,t) = \int c(x, t')\, dt'$, i.e., $\xi = \xi(D)$. In its turn, the concentration field is found by a deterministic manner by means of the parameters of the sub-system $O \bigcup M$, that is, $c = c(p(t),m(t))$. That is, in this case the function of losses is a complex function from $p(t)$ and $m(t)$.

Do the following mental experiment. We will observe the system Σ over the interval $[0,T]$. Each observation of this type (outcome) $\omega_i \in \Omega$ will be characterized by its own values $p_i(t)$, $m_i(t)$ and, as follows from (1), $\xi_i(t)$. Hence, we come to a conclusion that $p = p(t,\omega)$, $m = m(t,\omega)$, $\xi = \xi(t,\omega)$ are random processes.

Determine risk as an expected value of losses ξ

$$\Re_{\xi}(t) \equiv < \xi(t,\omega) > . \qquad (2)$$

Expression (2) is the most generalized determination of risk. From this we come to an important conclusion that, in the most general case, the risk function is a time function, as ξ is a process.

To continue the presentation, it is necessary to build a sample space Ω, which is, generally speaking, an infinite dimensional. Represent Ω as

$$\Omega = \Omega^a \bigcup \Omega^s, \Omega^a \bigcap \Omega^s = \varnothing, \qquad (3)$$

where Ω^a is a finite space of accidental outcomes, Ω^s is a space of outcomes without accidents.

With the accepted notations the accidental outcome is an outcome described by the function $p^a(t)$ when at least one point $t' \in [0,T]$ is present

$$\lim_{t \to t'} \frac{dp^a}{dt} = \infty,$$

i.e., t' is, in fact, the moment of the accident. Usually a certain approximation (model) of the space Ω^a is determined as a result of analysis of the safety of the object O, then $\Omega^s = \Omega \setminus \Omega^a$. Taking into consideration (3), we get

$$\Re_{\xi}(t) \equiv \int_{\omega \in \Omega} P(\omega)\xi(t,\omega)d\omega = \int_{\omega \in \Omega^s} P(\omega)\xi(t,\omega)d\omega + \int_{\omega \in \Omega^a} P(\omega)\xi(t$$

$$= P(\Omega^s) \int_{\omega \in \Omega^s} P(\omega \mid \Omega^s)\xi(t,\omega)d\omega + P(\Omega^a) \int_{\omega \in \Omega^a} P(\omega \mid \Omega^a)\xi(t,\omega) \qquad (4)$$

$$= (1 - P(\Omega^a)) \cdot \Re_{\xi}^{s}(t) + P(\Omega^a) \cdot \Re_{\xi}^{a}(t),$$

where $\Re_{\xi}^{s}(t)$ is a systematic component of risk or the real risk (due to normal functioning of the object O without accidents), $\Re_{\xi}^{a}(t)$ is an accident component of risk or the potential risk. Note that, generally speaking, $P(\Omega^a) = P(\Omega^a, T)$, i.e., it depends on the prediction horizon *T*. Due to the fact that, normally, $P(\Omega^a) << 1$, the following approximation is true

$$\Re_{\xi}(t) \approx \Re^s_{\xi}(t) + P(\Omega^a) \cdot \Re^a_{\xi}(t). \tag{5}$$

Expression (5) is very general and cannot be applied directly to practical computation.

3. MODEL OF RISK PREDICTION

Do the following simplifications and assumptions. Consider the model of risk of the negative effects transmitted by means of the free air without restricting the community. That is, we will mean the planetary boundary layer by the environment *M*. Assume that the processes *p(t)* and *m(t)* are independent statistically. The process *m(t)* (in this particular case representing the behavior of meteorological elements) can be reproduced by the data of the climatic observations in the area of location of the object done over many years by means of statistic modeling [1].

Consider the systematic component of risk $\Re^s_{\xi}(t)$ is the expected value ξ by definition, in which connection the elementary events are represented by the specific realizations of the processes *p(t)* and *m(t)*. From here

$$\Re^s_{\xi}(t) = \int\limits_{p_0}\int\limits_{m_0} \rho(p_0)\rho(m_0)\overline{\xi}(t, p_0, m_0)dp_0 dm_0, \tag{6}$$

where p_0, m_0 are the initial conditions for the processes p and m, respectively; $\rho(p_0)$, $\rho(m_0)$ are the density of distribution of the probability of realization of the initial conditions,

$$\overline{\xi}(t, p_0, m_0) = \lim_{N \to \infty} \frac{1}{N} \sum_{i=1}^{N} \xi(p_i(t), m_i(t) \mid p_i(0) = p_0, m_i(0) = m_{\text{(}} \tag{7}$$

In the latter expression i is the number of combined occurences of the random processes p and m. In the case of long-term action it is possible to expect that $\overline{\xi}$ will not depend on the initial conditions, i.e., the following approximation is true

$$\Re^s_\xi(t) \approx \bar{\xi}(t, p^*, m^*), \tag{8}$$

where p^* and m^* are the arbitrarily selected and physically realistic initial conditions in the capacity of which the most probable ones can be selected.

Begin considering the model of the accidental component of risk. Assume that there exists a finite mutually incompatible breakdown of the space Ω^a into the classes of accidents or scenarios C_z, z is the class number. Then

$$\Re^a_\xi(t) = \sum_z P(C_z \mid \Omega^a) \cdot \Re^z_\xi(t),$$

where $P(C_z \mid \Omega^a)$ is the relative probability of the scenario z,

$$\Re^z_\xi(t) = \int_{\omega \in C_z} P(\omega \mid C_z) \xi(t, \omega) d\omega, \tag{9}$$

the bottom line, the relative accident risk of the scenario z.

Further, the function $p(t)$ can be represented as $p(t) = p^s(t) + p^z(t - t')$, where p^s and p^z are the systematical and accidental components, respectively. Note that $p^z(t - t') = 0$ at $t - t' \leq 0$. As a rule, $p^z >> p^s$, from this $p(t) \approx p^z(t - t')$. Hence, the function ξ for the scenario C_z will have the view $\xi = \xi(p^z(t - t'), m(t - t'), \omega)$, i.e., $\xi = \xi(\tau, \omega)$, $\tau \in [0, T]$, where $\tau = t - t'$ is the time calculated from the moment of the accident.

If we assume further that the moment of the accident is distributed uniformly over the prediction horizon $[0, T]$, we will get the following

$$\Re_{\xi}^{z}(t)=\int_{t'=0}^{T}\rho(t')\cdot\Re_{\xi}^{z}(t-t')dt'=\frac{1}{T}\int_{0}^{t}\Re_{\xi}^{z}(\tau)d\tau, \tag{10}$$

$$\Re_{\xi}^{z}(\tau)=\int_{m_0}\rho(m_0\mid C_z)\xi^{z}(\tau,m_0)dm_0, \tag{11}$$

$$\xi^{z}(\tau,m_0)=\lim_{N\to\infty}\frac{1}{N}\sum_{i=1}^{N}\xi(p^{z}(\tau),m_i(\tau)\mid m_i(0)=m_0), \tag{12}$$

where formula (10) corresponds to the averaging over time from the beginning of the accident, formula (11) over parameters of the environment at the moment of the accident, formula (12) over various occurences of the process $m(\tau)$ at the initial conditions set at the moment of the accident.

Note the important consequences from (9)-(12).

1. In analyzing relative risks $\Re_{\xi}^{z}(\tau)$, it is necessary to consider not only the relative probabilities $\rho(m_0 \mid C_z)$ of occurence of the meteorological elements at the moment of the accident, but also the process of change of the meteorological elements $m(\tau)$ during the subsequent moments of time.
2. Under the assumption that the process $m(\tau)=m_0=const$, we will get the well known approximation

$$\Re_{\xi}^{z}(\tau)=\int_{m_0}\rho(m_0\mid C_z)\xi(p^{z}(\tau),m_0)dm_0. \tag{13}$$

3. The relative risk $\Re_{\xi}^{z}(\tau)$ is a function of only one variable τ that is the time interval from the moment of the beginning of the accident.
4. Since $\lim_{\tau\to\infty}\Re_{\xi}^{z}(\tau)=\Re_{\xi}^{z}(\infty)=const$, the following upper bound is true

$$\Re_{\xi}^{z}(t) \leq \Re_{\xi}^{z}(\infty)\frac{t}{T}, \quad \Re_{\xi}^{z}(T) \leq \Re_{\xi}^{z}(\infty). \qquad (14)$$

4. CONCLUSION

This work presents an attempt of systematical approach to the problem of risk assessment. A formal model of risk prediction comprising real and potential risks is built on the basis of an analysis of the unsteady process of influence of the chemically dangerous facility on the environment. Note that the proposed model is not referred to any specific choice of the damage function or a specific mechanism of creation of the space of accident events. It is possible to obtain a specific model of risk prediction adequate for practical computation by concretizing the damage function ξ, the sample space Ω^{a} and the process *p(t)*. The proposed model can be used as a formal basis for thorough study and comparative assessments of the environmental risks associated with disposal of chemically dangerous compounds.

ACKNOWLEDGMENT

The work was made possible thanks to the financial support of the Russian Foundation of Basic Research (Grant 01-01-96444) and Research Programme "Russian Universities" (Grant УР.03.01.015).

References

[1] Kisselev V.B., Kouznetsov V.N., Miljaev V.B. (1999). A comparison of mesoscale pollution transport models with varying spatial. In: R. San Jose, ed. Measuring and Modelling Investigation of Environmental (Vol 2). Southampton: Computational Mechanics Publications, pp. 187-204.

THE COMPLEX APPROACH IN APPRAISAL OF THE CONDITIONS OF NATURAL ECOSYSTEMS IN THE «VALLEY» ZONE USED FOR THE STORAGE OF CHEMICAL WEAPONS (POCHEP, BRYANSK REGION)

V. P. Ivanov[1], I. A. Baljasnicov[2], V. P. Sheluho[1], I. N. Glazun[1], D. I. Nartov[1]
[1]Bryansk State Academy of Engineering and Technology, Bryansk (Russia), [2]The Bryansk Regional Authority of Natural Resources and Environmental Protection, Bryansk (Russia).

Abstract: Research was carried out on the ecological situation in 2001 on the territory of the largest arsenal of chemical weapons, in the Pochep region of Bryansk area. These chemical weapons are an agent dispersal ammunition, which is filled with nerve agent. The complex method was used in the analysis of natural ecosystems. Bioindicators of the state of environment were chosen among the components of forest ecosystems such as: standing timber of a Scotch pine (Pinus silvestris L.); fir-tree (Picea abies (L.) Karst), oak (Quercus robur L.), regrowth, undergrowth, topsoil growth, soil mesofauna, ornithofauna, amphibians, assemblage of epiphyllous lichen, the pollen of a Scotch pine and also agrocenosises. The system of local ecological monitoring was established on the territory of the arsenal «Valley». The system is based on 23 permanent trial areas. On the whole, no negative influence of the arsenal of chemical weapons on the components of natural ecosystems was revealed. Marked negative changes of separate indices are caused by the influence of additional technogenic and anthropogenic factors.

Key words: monitoring, bioindicator, regrowth, undergrowth, soil

In the Pochep region of the Bryansk area, in the huge tracts of forest there is the largest arsenal of chemical weapons. It is the agent dispersal ammunition, which is filled with nerve agent such as sarin, soman, VX-gases. The allocation of chemical weapons to the arsenal, which is called «Valley», began from the end of the 1960s. The chemical ammunition is kept without fuses. All storehouses are equipped with the system of fire-prevention security. There were no emergencies in the «Valley» during the storage of the air chemical ammunition. Bad ecological conditions in much of the territory of Bryansk after the crash on the «Chernobyl» nuclear power station disturbed the inhabitants of Bryansk region. [1]. Information about the large arsenal of chemical weapons in Pochep region and about its pending processing has increased the stress in the population. The carrying

V.M. Kolodkin and W. Ruck (eds.),
Ecological Risks Associated with the Destruction of Chemical Weapons, 99–105.

out of the given research is the prominent aspect of protection of environment, along with the storage and destruction of chemical weapons.

The analysis of components of nature gives the real picture of the states of ecosystems at the current moment. The huge tracts of forest are perfect bioindicators of quality of the environment. They give the opportunity to monitor all levels of organization of living matter [2].

The research of ecological situation in the zone of the chemical weapons arsenal «Valley» were carried out under the initiative of the Committee of Natural Resources in the Bryansk area in 2001. The program of research and development operations provided the comprehensive method of analysis of ecosystems.

The following forest components were selected as bioindicators of control of the state of the environment: standing timbers of a Scotch pine (Pinus silvestris L.); fir-tree (Picea abies (L.) Karst), oak (Quercus robur L.), regrowth undergrowth, topsoil growth, soil mesofauna, ornithofauna, amphibians, assemblage of epiphyllous lichen, the pollen of Scotch pine and also agrocenosises of adjoining territories.

The research consists of:

- analysis of structure of the forest resources;
- study of biological variety of phytocenosises;
- preliminary inspections of standing timbers and detailed inspection of 23 permanent trial areas (P.T.A), with number of trees not less than 100 pieces on one plot with predominance of pine, fir-tree, oak, both in the researched region and in the control zone (100 km from the arsenal);
- measurement of the indices of exposure (diameter at height 1,3 m for all trees on P.T.A.; height of 20-25 trees);
- sampling of wood of 15 trees on every P.T.A. for performing dendrochronological analysis;
- estimation of pathological state of forests and estimation of the role of pathological factors in the process of their weakening;
- analysis of the state of regrowth on P.T.A. and analysis of perspectives of the change of standing timbers;
- analysis of reproductive indices of Scotch pine (the research of pollen);
- analysis of lichenometrical and lichenoindication characteristics of lichen senysias (with accounting for 25 trees on every P.T.A.);
- analysis of structure and prevalence and the occurrence of life forms of soil mesofauna (by digging up 12 pits on every P.T.A.);
- the study of variety of species of birds;

- revelation of variety of species and morphological variability of amphibians;
- ground research;
- analysis of the state of agrolandscape.

Such a wide set of natural components, which includes the characteristics of their state, is used for the 1st time for the analysis of the ecological situation in such zones.

The research was carried out by applying methods which are used for the analysis of landscape, soil and exposure and pathological characteristics of forests. The use of modern methods allowed to study in details the group of factors, which affect the state of components of natural ecosystems on the territory of arsenal «Valley».

The resulting research of this wide set of the most sensitive components of biogeocenoses revealed a negative reaction only in some of them, in the 2nd km zone around the arsenal «Valley», which is a flourishing facility of anthropogenic character. It has been in existence for several decades.

The sanitary and pathological state of forests depends on number of ecological factors and bioecological parameters of standing timbers. The worst indices of the sanitary state of forests are marked on the P.T.A., which are situated near (up to 20 km) the arsenal of chemical weapons. The forests with broken biological stability (2nd class) dominate. 66.5% of them are pine and oak forests, 80% of them are spruce forests. The others belong to the 1st class of biological stability.

Marked lowering of the category of state and biological stability of oak-groves in quarters 4 and 5 of Semetskoe forestry and in the 53rd quarter of Bryansk military forestry may be connected to the influence of the sources of technogenic pollution of Pochep.

The lowering of radial increase intensity is typical for all studied trees (up to 40%) from 1981 up to 2000. The oaks and pines have lower rates of increase in the short range zone (up to 0.6 km from perimeter of the arsenal). Such dependence is not marked for spruce. The negative tendency of dependence of radial increase on the distance up to the perimeter of the facility is explained by smaller recreational loads because of the special regime status of adjoining forests and restricted fellings of forests in Bryansk military forestry. The analysis of even-aged trees (55-60 years old) has shown the negative influence of the arsenal on radial increase of the oaks but there is no such regularity in pine forests. The essential influence on the dynamic of radial increase of the trees is exerted by climatic factors. However, in coniferous forests of a short range zone, the regular dependence of current radial increase on climatic indices is marked with less number of parameters. It is explained by influence of additional anthropogenic factors such as the road «Pochep-Pogar», the nearness of technogenic facilities of Pochep, and the military town.

The study of the regrowth and undergrowth points out the renewal of natural standing timbers on the studied territory. The renewal is better in fir-groves but it is worse in pine-groves. It depends on the age and density of standing timbers and it is also depends on the dampening action of well-developed undergrowth. The direct dependence of the density of the regrowth of oak-groves and fir-groves at a distance from the arsenal was found. In moving away, the density of reliable regrowth increases and the density of doubtful regrowth decreases.

Insignificant increase in the density of the regrowth up to 2 km from the arsenal on some P.T.A. is caused by lessening of the competition from the side of maternal tier because it is weakening. The dependence of sanitary state, average increment of height of the regrowth and undergrowth on the distance up to the arsenal is not evident.

Not only such factors as acidity and fertility of the soil, structure and the age of standing timbers but also the anthropogenic factor (probably, the military town which is included in the arsenal) affects the condition of grassy vegetation on the studied area. It caused the lowering of the general number of species and number of predominant species while approaching the perimeter of the facility. Often predominant species on distant P.T.A. end up being the single one on P.T.A. which are near the arsenal.

The change of parameters of lichenosynusias on standing timbers of fir, pine, and oak, nearer the the perimeter of the arsenal, points out its negative influence on the adjoining territories (up to 2 km) though the level of influence is close to the standard.

Bioindicational research with use of cytological methods for analysis of pollen according to such indices as ability to fecundation, vitality, the length of pollen tube, and the size and shape of pollen seeds have not revealed a negative influence of the arsenal on the reproductive capacity of a pine up to 3 km. However, the number of anomalies in the ability of fecundation are a little higher near the perimeter of the facility (up to 0.3 km). The ecological conditions were favorable for the reproductive process of trees at the moment of research.

During the study of soil mesofauna on P.T.A. the lowering of average density of settlement of phytophages was marked in fir groves while moving as the distance from the facility increased. In pine groves this index is close to the standard conditions on different distances. In oak groves and fir-groves the density of phytophages and zoophages is identical: the maximal index of density of settlement is marked on the territory up to 5-6 km from the arsenal. The average density of suprophages in standing timbers of all type falls with increasing distance from the boundaries of the facility up to the controlled territory. It is caused by the increase in the number of litter-fall in the forests, which are near the facility (up to 3 km).

In the future the usage of species variety and characteristics of populations of zoophages and saprophages, as having the greater importence, is more expedient for monitoring the impact of the arsenal on the environment. Also it is possible to use the species variety of phytophages, the density of the settlement, and the occurrence of grubs of Elateridae.

The variability of amphibians (for example frogs) on the polluted territories shows high departure from the norm of the number of characteristics. It is regarded as an indicator of the state of the environment. The fluctuating asymmetry or frequency of occurrence of asymmetric signs on the territory which is near to the facility «Valley» is equal to 0.312. On the adjoining territory (approximetly 4 km from the facility) the frequency of occurrence of asymmetric signs is equal to 0.302 (the lake near the road Pochep-Semtsi) and the frequency of occurrence of asymmetric signs is 0.266 on rather clean territory (the flood land of Rozok river, approximately 6 km from the arsenal). These results show that amphibians can be indicators of the state of environment and reveal of pollution intensity, including chemical pollution.

The analysis of species variety and of the state of ornithofauna testifies that , as a rule, forests, which contain fir exert a positive influence on the increase of the number of settlements of birds and species abundance. The researched region is characterized by a wide variety of birds (62 species). It is interesting that 51 birds species live in these forests. On the whole, the number of wild birds is 4.6 times higher than on the open area.

The change of species variety and density of settlement of birds is not, probably, connected to the influence of arsenal «Valley». The special regime of the restricted area produces the favorable conditions for nesting, reproduction, and increase of species variety of birds. The system of monitoring will help to do the study of dynamics of ornithofauna in more detail.

The tendency towards degradation of agricultural soils is pronounced on the studied territory. It is explained by economic conditions of the country for the last decade. Such research with analysis of a great number of components of nature was carried out on the territory of the arsenal «Valley» for the first time. The real situation of the state of the natural ecosystem was revealed. We can make judgements about the direction of processes on the studied territory only by separate parameters, such as the increase of diameter of the trees, the character of natural renewal, the dynamics of forest resources for the last years. It is expedient to create a system of monitoring on the basis of permanent trial areas, which were founded by a scientific group of employees of Bryansk State Academy of Engineering and Technology.

The created system of local monitoring will help to find out the influence of different technogenic and anthropogenic sources of pollution of Pochep and villages and collective farms on the adjoining areas. The regular observations, the comparison of obtained characteristics with the analogy of P.T.A. on the control zone will help to find out the direction of the ecological process, and they will give the opportunity to make prognoses of the ecological situation and to elaborate the recommendations for its stabilization in case of a worsening of the ecological situation.

References

[1] Report on the State of the Environment of Bryansk Region in1999 and 2000.
[2] Alekseev, A.S. (1996). Theoretical Basis of Population of Bioindication of Anthropogenic Influence of Forest Ecosystems. The Influence of Atmospheric and Natural Factors on Destabilization of the State of the Forests in Central and East Europe. MSUL, M, Vol 2, p. 103-106.

ANALYSIS OF ECOLOGICAL RISKS WITHIN AN ENVIRONMENTAL MONITORING SYSTEM

Nicolay Zabrodine[1], Alexander Churakov[2], Ioulia Bouchmakina[1], Sverre Langard[3]

[1]The Centre for State Sanitary and Epidemiological Inspection in the Udmurt Republic, Izhevsk (Russia), [2]The Izhevsk State Medical Academy, Izhevsk, (Russia), [3] The Centre for Occupational and Environmental Medicine, The National Hospital, Oslo (Norway)

Abstract: Environmental monitoring is a system of observation, estimation and prognosis for the environment under the influence of natural and anthropogenic factors. This system must be realized in the frames of 4 sub-systems, which are air monitoring, surface and ground water monitoring, soil monitoring, and flora and fauna monitoring. The effectiveness of the environmental monitoring, the realization of its purposes and tasks depends on the cooperation between "the informative sub-system" and the health care institutions which are responsible both for the social-hygienic and the social-ecological monitoring, the other organizations, and the administrative and governmental institutions.

Key words: ecological risk, environmental monitoring, health care institutions

1. INTRODUCTION

Environmental monitoring is a system of observation, estimation and prognosis for the environment under the influence of natural and anthropogenic factors (in the terms of the "Russian Federal Environment Protection Act"). The technologies for environmental monitoring consist of the tools and techniques for observation, and the creation and estimation of the recommendations for the purpose of environmental management. The recommendations and management include the estimation of indexes for preservation and prognosis of the state of an environment, the description of ecological and technological characteristics of occupational conditions, and the evaluation of the medical, biological and sanitary-hygienic conditions of the existence of man and biota. Social-hygienic monitoring is a part of environmental monitoring. It is a state system for an observation analysis, evaluation and prognosis of both the environment and the state of health of the population. This state system includes the estimation of the influence of

V.M. Kolodkin and W. Ruck (eds.),
Ecological Risks Associated with the Destruction of Chemical Weapons, 107–110.

environmental conditions on the health of the population and social-ecological monitoring, which is the system of observation of public opinion about environmental problems.

The public reaction to the changes within the environment is a topical problem. The public unease arises from both the non-stable economic situation within the country and the negative influence of an environment. This public unease is a typical reaction among people who are living in "the territory of ecological risk". Thus, the activities for environmental protection during the construction and operation of the possibly dangerous "facility" need a frank discussion. The explanations given to the public about the purposes of the construction and operation of the "facility" will lower the public unease and facilitate the management.

2. THE SYSTEM OF ENVIRONMENTAL MONITORING

Monitoring for environment protection must be performed within the frames of 4 sub-systems, which are air monitoring, surface and ground water monitoring, soil monitoring, and flora and fauna monitoring. The monitoring of each sub-system is based on the completion of the following management steps: the determination and investigation of a 'facility' for observation; the creation of 'an informational model' of 'the facility'; the planning of the observation and measurements; the organization and establishment of observation and measurement; the data collection and analysis; the evaluation of the state of the "facility" and its "informational model".

The main and most important principles for all sub-systems are the validity of results from the observation and measurements, the usage of devices and equipment that are in accordance with state requirements, the usage of certified programs, the analysis of the samples in accredited laboratories, and the documentation of procedures of analysis and measurements.

The organizational measures should be oriented on the accomplishment of the following tasks: monitoring; the creation of a data bank and data analysis; the estimation and prognosis of the state of the environment and the influence of the environment on the health of the population; the availability of an efficient data submittal and retrieval in the case of disaster with the follow-up prognosis of state of the environment and the damage; and the presentation of necessary information to the public.

It is necessary to subdivide the environmental monitoring into 2 blocks which are "the expert judgements" and "the informational block".

3. CONCLUSIONS

The effectiveness of environmental monitoring, and the realization of its purposes and tasks depends on the cooperation between "the informative sub-system" and the health care institutions, which are responsible both for the social-hygienic and the social-ecological monitoring, the other organizations, and the administrative and government institutions.

References

[1] Birukova R.N., Dogle N.V., Sluchanko I.S. (1966). Sanitary statistic. Moscow.
[2] Onishenko G.G., Novikov S.M., Rachmanin Yu,A. (2002). The risk assessment for the population health. Moscow.
[3] World Health Organization. (2000). The air monitoring due to influence of man's health. Copenhagen.
[4] Dautov F.F. (1990). The influence of environmental factors on the population health. Kazan.

DECHLORINATION OF RECALCITRANT POLYCHLORINATED CONTAMINANTS USING BALL MILLING

Volker Birke[1], Jörg Mattik[1], Dietlind Runne[1], Helmut Benning[2], Dragan Zlatovic[3]

[1]University of Applied Sciences of North East Lower Saxony, Suderburg, Germany, Faculty of Civil Engineering (Water and Environmental Management), Steinweg 4, 30989 Gehrden, Germany, voice +49 5108 9217-30, birke@fhnon.de; [2]German Institute of Rubber Technology, Eupener Straße 33, 30519 Hannover; [3]L+Z Entsorgungsdienste für Starkstromanlagen GmbH, Am Bubenpfad 2, 67065 Ludwigshafen am Rhein, Germany

Abstract: The persistence of hazardous pollutants in the environment or waste streams and materials is frequently due to the thermodynamic and/or kinetic stability related to their molecular structure. Further these pollutants tend to be sorbed strongly to other components of the contaminated matter such as polychlorinated biphenyls (PCBs) or pentachlorophenol (PCP) to clayish soils or polychlorinated dibenzodioxins (PCDDs) or dibenzofuranes (PCDFs) to filter dusts. Adsorption renders such pollutants virtually inaccessible to physical, chemical or biological remediation and/or natural attenuation processes. These difficulties of stability and inaccessibility can be effectively overcome by mechanical activation of appropriate reductive dehalogenation reactions inside a *ball mill*. Such recalcitrant compounds, like PCBs, PCDDs, PCDFs or PCP, can be completely dechlorinated under strikingly benign, reductive conditions, i.e., *at room temperature*, *ambient pressure* and *in a short time*, by *ball milling* the contaminated material with a base metal and a hydrogen donor. A ball mill is utilized as a mechanochemical dehalogenation reactor in one single, almost universally applicable, operational key step. This novel, versatile *ex situ* dehalogenation approach can be successfully applied both to contaminated materials and highly concentrated or pure contaminants and of their mixtures, *virtually regardless of their state*, and is designated as "Dehalogenation by Mechanochemical Reaction" (DMCR). For instance, PCBs in contaminated soil, filter dust, transformer oil, or as pure PCBs oils from transformers or capacitors can be rapidly dechlorinated to harmless chloride and their parent hydrocarbon biphenyl by applying magnesium, aluminum or sodium metal and a low acidic hydrogen source like an ether, alcohol and/or amine. DMCR offers several economic and ecological benefits. Ball milling requires a low energy input only, toxic compounds can be converted to defined and usable products, the method facilitates the re-use of scrap metals such as magnesium or magnesium-aluminum alloys, and detoxified materials like transformer oils can be readily recycled after a DMCR treatment. No harmful emissions to the environment have to be expected. This paper presents selected results of various treatability and feasibility studies demonstrating the versatility and efficiency of DMCR regarding the defined destruction of numerous hazardous persistent organic pollutants (POPs) like

V.M. Kolodkin and W. Ruck (eds.),
Ecological Risks Associated with the Destruction of Chemical Weapons, 111–127.

PCBs, PCDD/Fs and related compounds (PCP and dichlorobenzene (DCB)) in different solid and liquid contaminated materials.

Key words; Mechanochemistry, mechanochemical, dehalogenation, dechlorination, magnesium, waste, disposal, destruction, PCBs, PCDD, dioxin, DDT, pesticide, mustard gas, lost, lewisite, POPs, ball mill, milling, grinding

1. INTRODUCTION

Chemistry occurring at surfaces under impact has been investigated for a long time and has found numerous practical applications. Such phenomena are commonly assigned to two branches of chemistry, i.e., "mechanochemistry" and/or "tribochemistry". Mechanochemistry deals with the chemical and physico-chemical changes of *substances of all states of aggregation* due to the influence of mechanical energy. Tribochemistry deals with chemical and physico-chemical changes of *solids* due the influence of mechanical energy (Heinicke, 1984).

Recently, it was shown that polyhalogenated pollutants can be destroyed by mechanochemical reactions. This opened up the development of novel, innovative *ex situ* remediation and decontamination processes (Rowlands et al., 1994; Donecker et al., 1994; Hall et al., 1996; Birke, 1996, 1998, 2001, 2002; Loiselle et al., 1997; Cao et al., 1999; Gock et al., 2001; Ikoma et al., 2001; Aresta et al., 2002, 2003). Mechanochemical degradation can be easily performed using *ball mills* that are readily available in different sizes (treatment of materials up to several tons is possible) and constructions. The pollutants are eliminated directly inside a contaminated material, virtually regardless of how complex its structure may be and how strongly the pollutants may be bound adsorptively to particular compounds (e.g., clays regarding soils). Usually, mechanochemical degradation can be performed at low temperatures, i.e., room temperature. Inside the material to be milled, however, very high temperatures up to several thousand degrees Celsius at points of action in the sub-microscopic deformation zone, where a grain collides at high velocity with a solid surface, may occur (quasi-adiabatic energy accumulation, so-called "triboplasma" (Heinicke, 1984)).

However, none of today´s known mechanochemical approaches for the destruction of hazardous polyhalogenated pollutants has been scaled up to an industrial scope or even commercialized so far, most probably due to two major reasons. First, some approaches generate/leave numerous intermediates and degradation products, still partly halogenated, associated with an unknown, but evidently potential toxicological risk, e.g., during milling of DDT with calcium oxide (Rowlands et al., 1994, Donecker et al., 1994, Hall et al., 1996). For others, actual degradation products have not been identified at all yet. Second, reaction times predominantly range between several hours or even days, therefore revealing a considerable disadvantage compared to common, well-established *ex situ* destruction techniques like incineration. However, a recently developed approach using certain hydrides may be promising regarding scaling up for several fields of application (Aresta et al., 2002, 2003).

Here, a novel mechanochemical dehalogenation approach for the destruction of recalcitrant polyhalogenated compounds like PCBs is reported, using simple and readily available reagents, supplying defined degradation products and showing the potential to afford reaction times in the range of minutes to only a few hours. The paper's focus is on reductive dechlorinations of polychlorinated aromatic compounds, because these are especially recalcitrant, and some of them represent prominent POPs. For that purpose, base metals like alkali and alkaline-earth metals, but also aluminum, zinc and iron, can be deployed in combination with certain hydrogen donors like alcohols, ethers, amines, and so forth. Figure 1 outlines the general DMCR approach for PCBs dehalogenated by magnesium metal in a ball mill, as one of numerous combinations/examples conceivable/possible.

MILLING
(RT, min-hrs)

R–*Cl*	+	Mg	+	"*H*"	⇒	R–*H*	+	MgCl_2
(PCB)		*Base Metal (Magnesium)*		*Hydrogen-Donor*		*Biphenyl*		*Salt*

TOXIC **LESS TOXIC/ HARMLESS**

Figure 1: Overview of the DMCR method. One example outlined (rough, simplified scheme, not stoichiometric): during ball milling, PCBs are reductively dechlorinated by magnesium metal in the presence of appropriate hydrogen donors (alcohols, ethers, amines, and so forth) at room temperature yielding biphenyl and magnesium chloride.

Below, examples of basic mechanochemical reductive dechlorinations of PCBs, PCP, PCDD/Fs and 1,3-DCB by milling with magnesium or sodium are displayed and discussed, which have been part of various laboratory studies and pilot projects covering contaminated soils (Birke, 2000 and

2001), oils, fly ashes and solid or liquid waste. Evidently, DMCR treatment can be successfully applied to concentrations spanning from ng/kg up to even pure contaminants.

2. MATERIALS AND METHODS

Centrifugal ball mills ("Retsch S 1" and "Retsch S 1000", supplier "F. Kurt Retsch GmbH & Co. KG", Haan, Germany), each equipped with a 50 ml steel grinding jar (containing three steel balls, diameter 2 cm), were used for bench-scale treatability studies. An "ESM 234" vibratory mill (SIEBTECHNIK GmbH, Mülheim an der Ruhr, Germany) (TRIBOCHEM, 2003) was utilized for testing dechlorination of PCBs in transformer oils at a kg scale. The mill was equipped with inner steel lining and loaded with steel balls (diameter appr. 2 cm each), up to/consuming 80% of the whole volume capacity. Several slight modifications were mounted in order to adjust the device to some specific requirements during DMCR runs. In any run, the mill was loaded with the defined volume of PCB transformer oil, followed by adding the metal (sodium lumps or dispersion or magnesium filings) and finally the amine.

Dechlorination of PCBs/PCP. The 50 ml steel grinding jar was charged with sand, a mixture of calcium oxide and calcium hydroxide, and triethylene glycol dimethyl ether (triglyme). Pure Chlophen A 30 was added, and the materials were blended by grinding for one minute (Chlophen A 30 is a former PCB brand of BAYER GROUP, Germany, representing a technical PCB mixture containing 42 % chlorine (w/w), which equals approximately to a trichlorobiphenyl congener). Afterwards, the jar was loaded with methanol and magnesium. The milling process was started and maintained for the required reaction time at about 530 rpm ("Retsch S1" or "Retsch S 1000"). After that, the finely powdered mixture was extracted with n-hexane for 20 hours, employing a Soxhlet extractor (prior to the extraction it was carefully mixed with methanol in order to destroy excessive magnesium metal). It was verified that no additional dechlorination had occurred during work-up. The n-hexane phase was washed with water and nitric acid. Then, its volume was reduced to a few milliliters by carefully evaporating at ambient pressure in a Kuderna-Danish-Evaporator. After having cleaned up the reduced n-hexane volume over a small silica gel column and adjusted it to a defined volume, an aliquot part was injected to a GC MS line (Hewlett Packard, model 5972, split/splitless injector, injection according to Grob´s method, i.e., splitless during the first 1.5 minutes, afterwards with opened split (1 : 23), "HP-5" column (phenyl-methylsilicon), length 30 m, carrier gas: helium; program 1 min. isothermal at 40 °C, ramp 15 °C/min. up to 280 °C).

In case of PCP, the 50 ml steel grinding jar was loaded with 15 g of sand containing appr. 4,300 ppm PCP (appr. 99 % pure, Merck), 0.76-1.5 g magnesium filings and 0.32-0.64 ml ethylenediamine. The milling process was started and maintained for 30 minutes at about 530 rpm. After that, an aliquot of 6 g of the finely powdered mixture was transferred into a flask and mixed with 15 ml water and 10 ml 6 N hydrochloric acid while being cooled down with ice. 10 ml methyl-tert.-butylether (MTBE) were added and the mixture was heated to reflux under vigorous stirring for 30 minutes. Afterwards, the organic phase was separated, purified by centrifugalizing and filtering and dried (over sodium sulfate). An aliquot part was injected to a GC FID line.

Time-dependent, isothermal dechlorination of 1,3-DCB in a sand slurry. The 50 ml steel grinding jar was loaded with 10 g sand, 1.5 g magnesium filings and 15 ml of a mixture of DCB and n-butylamine dissolved in cyclohexane (0.0389 ml DCB, 0.5 ml n-butylamine). The mill ("Retsch S 1000") was operated for the required time isothermally at the required temperature by deploying an appropriate external cooling/heating jacket (coolant/heating liquid: water), fitting tightly to the steel grinding jar. After ceasing the milling process, the slurry was immediately removed from the jar and filtered. Three ml of the resulting liquid was treated with acidified water and washed with 10 ml n-hexane in a separation funnel. The organic phase was dried over sodium sulfate and afterwards adjusted to a defined volume by adding n-hexane. A small aliquot was injected into a GC FID line. To get reliable data, for each data point the described procedure was entirely repeated from the start and was conducted in another, identical 50 ml steel grinding jar, too. The majority of the experiments was conducted in duplicate or triplicate for each given/selected data point, i.e., time value.

3. RESULTS AND DISCUSSION

Dechlorination of PCBs in solid matter and transformer oil, and of pure PCBs. According to Table 1 and the corresponding Figures 2-4, reductive dechlorinations of PCBs to biphenyl occur during ball milling at *room temperature*, i.e., when grinding a PCB/sand mixture with magnesium metal in the presence of methanol/triglyme. The extent of the dechlorination depends on several parameters. Specific operational parameters of the mill such as rotational speed, kind, size and number of balls were held constant during these studies, so that the milling time and the amount of the dehalogenating agent become the critical parameters (Table 1).

Investigations regarding reaction kinetics suggest that the involved mechanisms seem to be identical in principle to those identified in solution. In solution, alkali metals such as sodium reductively dehalogenate halogenated aromatic compounds in the presence of hydrogen sources like alcohols or ethers via radical or organometallic intermediates (Figure 5). Studies are currently under way to elucidate single steps in detail.

Time-resolved concentration measurements of the mechanochemical dechlorination of 1,3-DCB (Figure 6 and 7, DCB was used as a model substance for a PCB congener, results according to GC FID analyses) show – regardless of whether the metal was pre-milled/activated or not – that a specific lag phase occurs, followed by a rapid decay of the compound within 10-30 minutes.

Table 1: Dechlorination of appr. 7,000 ppm PCB (Chlophen A 30) in sand by milling with magnesium and methanol/triglyme in a centrifugal ball mill "Retsch S 1" (GC MS analyses, external standard).

Mg (eq.)[a)]	23	16	12	12	5
$\xi_{Mg/soil}$,	4.6,	3.4,	2.3,	2.3,	1.1,
$\xi_{RgTot/soil}$[a)]	10.3 %	9.0 %	7.9 %	7.9 %	6.8 %
Time [hr]	5	5	5	7.5	5
Dechlorination (GC-MS)	complete[b)]	complete[b)]	modest[c)]	significant[c)]	poor
PCB turnover [%]	97.2 [100][d)]	97.1 [100][d)]	28.4	49.9	20.2
Biphenyl yield [%]	78.6	75.2	3.1	7.4	1.8

a Magnesium equivalents (eq.) related to total organic chlorine (Chlophen A 30 = 42 % chlorine content (w/w)), ξ = ratio magnesium / (sand + PCBs ("soil")) (w / w), and reagents in total (RgTot) / (sand + PCBs) (w / w).

b Dominating final product: biphenyl; traces: phenylcyclohexane, halogen-free dimethylbiphenyls, quaterphenyls, see also Figure 2.

c Main products: partly dechlorinated PCB congeners (two monochlorobiphenyls).

d First value: reduction of peak area related to total peak area of PCBs prior to degradation (external standard). Second value (in brackets): in the retention range, where PCBs had entirely vanished, traces of entirely dechlorinated compounds were found, therefore a total PCB degradation/dechlorination is verified (see also Figure 2).

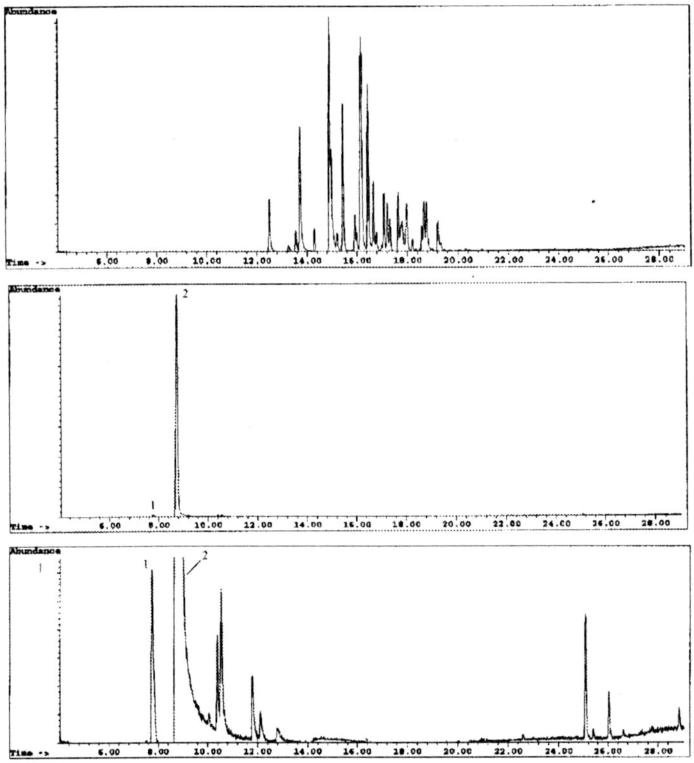

Figure 2: Complete dechlorination of Chlophen A 30 in sand (5 hrs) using 23 equivalents of magnesium (relative to organically bound chlorine) and methanol/triglyme: top = prior to treatment in a "Retsch S1" centrifugal ball mill, middle and bottom = after treatment (normal resolution, related to the GC on the top, and high resolution, resp.). Data see Table 1 (column #1). 1 = phenylcyclohexane, 2 = biphenyl, small peaks represent further, completely dechlorinated species (GC-MS analysis).

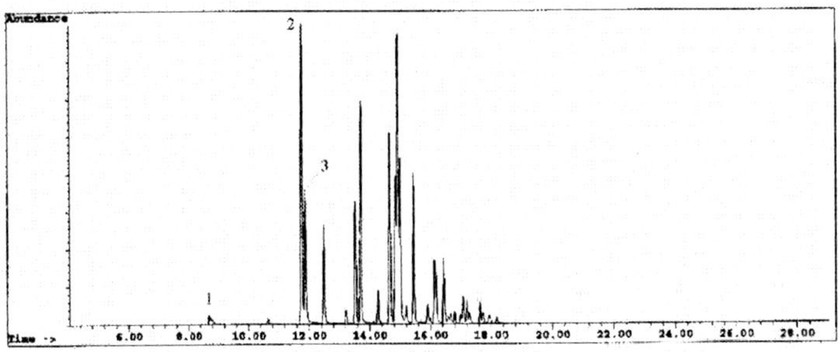

Figure 3: Incomplete dechlorination of Chlophen A 30 in sand during milling for 5 hrs with a reduced magnesium amount (12 equivalents of magnesium, see Table 1 (column #3)).
1 = biphenyl, 2, 3 = monochlorobiphenyls (GC-MS analysis), retention time [min.].

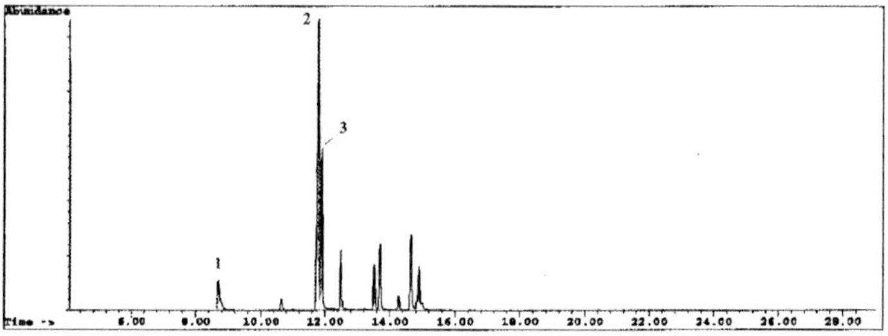

Figure 4: Significant, but incomplete dechlorination of Chlophen A 30 in sand during a prolonged milling time of 7.5 hrs, deploying again 12 equivalents of magnesium (see Table 1 (column #4)). Digits denote compounds being identical to those listed at Figure 3.

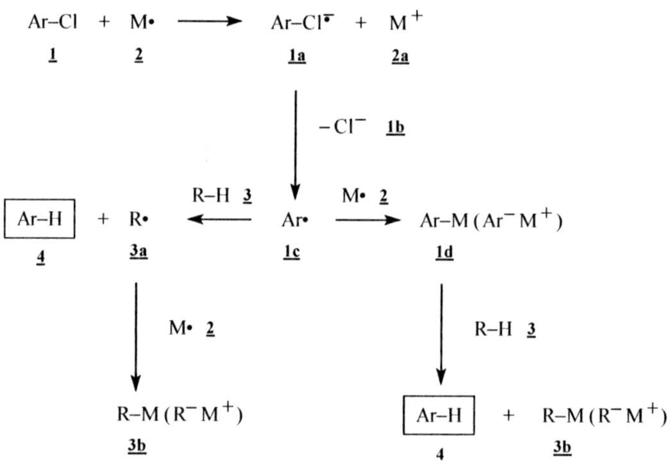

Figure 5: Reductive dechlorination of chlorinated aromatics (arylchlorides, Ar-Cl) by alkali metals in solution in the presence of hydrogen donors (R-H) like ethers or alcohols, compiled according to former publications (Smith and Ho, 1972 and 1973; Hawari, 1992).

The appearance of a lag phase (Figure 6 and 7) is qualitatively consistent with identical findings during other heterogeneous reactions performed in ball mills involving metals (Heinicke, 1984). Elevating slightly the temperature from room temperature to appr. 70 °C only reveals a significant influence on the course of the dechlorination: the lag phase is significantly reduced whilst the curve in the actual decay time window becomes steeper and steeper, i.e., the total time for the actual decay phase is reduced from appr. 20 minutes (at 25 °C) to appr. 10 minutes (at 68 °C) (Figure 7). However, pre-milling of magnesium seems to have not a major impact on the total time required for the dechlorination (see Figure 7 in comparison to Figure 6 representing runs without pre-milling/activating the magnesium at

25 °C). Temporary kinetic analyses of the data gained so far indicate that a specific Langmuir-Hinshelwood-like mechanism might govern the DMCR process. A process model will be set up in order to accurately design and adapt DMCR plants regarding every area of application.

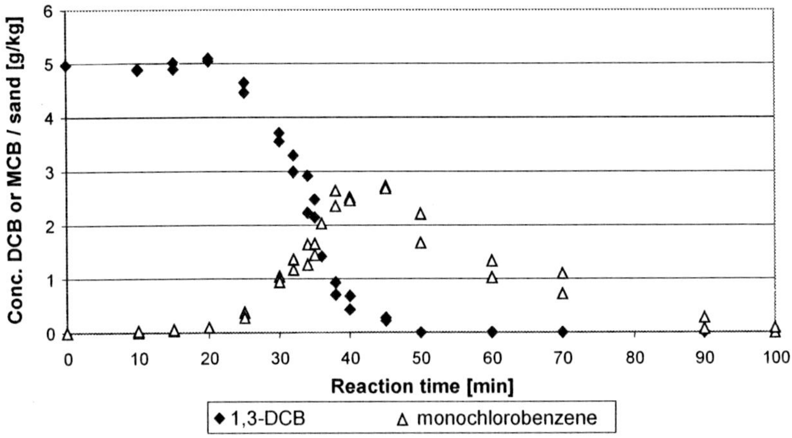

Figure 6: Complete reductive dechlorination of 1,3-DCB over time in a sand slurry with Mg, n-butylamine and cyclohexane (inert diluting agent) using a centrifugal ball mill (Retsch S 1000) at 25 °C. Benzene (not shown here) is the single and final dechlorination product.

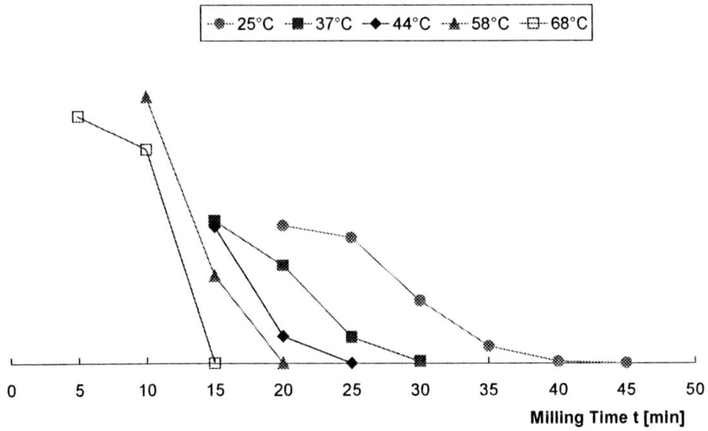

Figure 7: Temperature-dependent reductive dechlorination of 1,3-DCB over time in a sand slurry with pre-milled (pre-activated) magnesium, n-butylamine and cyclohexane using a Retsch S 1000 centrifugal ball mill (not to scale regarding the axis of ordinates; furthermore, DCB was employed in two different initial concentrations. Data represent average values of duplicate or triplicate runs (Janssen and Wille, 2002)).

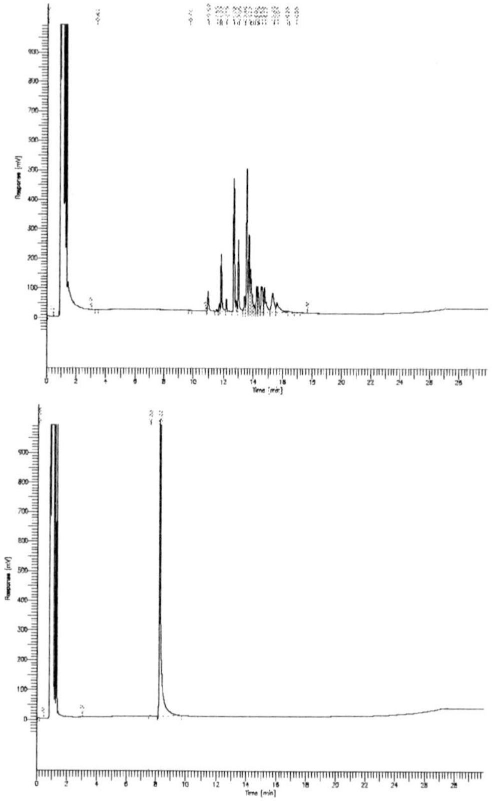

Figure 8: Accelerating the total dechlorination of Chlophen A 30 in sand: appr. 7,000 ppm PCBs (left) are completely converted to biphenyl (right) by appr. 23 equivalents magnesium (in comparison to Table 1, column #1) and *n-propylamine* during milling at room temperature *within 30 minutes*. Propylamine amounting to appr. 3 % (w/w) of the contaminated matter, i.e., instead of methanol/triglyme.

Accelerating the dechlorination. Lowering the reaction time required for a total dechlorination of PCBs towards the magnitude of minutes can be achieved by selecting other hydrogen donors, such as simple amines or amides instead of or in combination with alcohols and/or ethers (see Figure 8). This strategy can be successfully applied to both contaminated materials and virtually pure pollutants.

Pure PCBs (Chlophen A 30) could be entirely converted to biphenyl within *15 minutes* by using magnesium and amine (ratio PCB : Mg : amine = 1 : 1-2 : 3-5 (w/w/v)). Hence, DMCR makes it possible to get magnesium chloride and biphenyl in high yields, which can be used/sold as valuable chemical compounds. Biphenyl isolation demonstrated a 90-94 % recovery of biphenyl (mole/mole).

Amine-mediated reductive dechlorinations of DCB already shown above in Figure 6 and 7 represent reaction conditions, which were carried out with diluted reagents to increase reaction times and to expand/distend the time range for improved kinetic measurements/data observations. Under normal conditions, DCB can be completely dechlorinated to benzene within a few *minutes* only, if required.

According to Table 3, PCB contaminated transformer oils can be rapidly decontaminated *and recycled*, because the pollutants are destroyed under benign conditions *at room temperature*.

Table 2: Reductive dechlorination of PCBs with sodium or magnesium and n-propylamine in contaminated transformer oils at room temperature during milling in a modified SIEBTECHNIK "ESM 234" vibratory mill (according to German DIN standard, i.e., basing on standardized GC-ECD analyses performed in a certified lab).

Run #	PCB concentration [ppm]/ amount of transformer oil [L]	Metal [g] / % per oil	Time [min]	PCB conc., LAGA [ppm]
1	123.5 / 3	Na 10 g = 0.33%	5 10 20	30.0 17.5 **< 2.5**
2	123.5 / 3	Na 20 g = 0.67 %	1	**< 2.5**
3	308.5 / 3	Na 20 g = 0.67 %	2.5	**< 2.5**
4	123.5 / 2	Mg 200 g = 10 %	30 60 90	47.5 34.0 5.5
5	123.5 / 2	Mg 400 g = 20 %	10 20 30 40	28.5 17.5 2.5 **< 2.5**
6	123.5 / 3	Mg 150 g = 7.5 % + additive	30	**< 5.0**

As a result of the DMCR treatment, transformer oils remain intact, ready for re-use after purification: relevant analyses proved that the electrical properties of the oil can be guaranteed. Table 3 presents selected results of a study focusing in part on speeding up the dechlorination rate in the oil. The large excess of magnesium used can be further employed for numerous identical, tiered treatment steps in a semi-continuous process design. The low-boiling amine can be readily recovered and re-used. Similar findings were obtained for capacitor and used lubricating oils. By using DMCR, oils contaminated by polyhalogenated pollutants are not required to be incinerated any longer.

PCB dechlorination does not occur stoichiometrically. For example, a certain excess of magnesium is required to achieve complete dechlorinations (i.e., at least 16 equivalents related to organic bound chlorine according to Table 1). With sodium, at least seven equivalents in similar experiments are required. That observation can be explained by unavoidable side reaction in which biphenyl is further reduced by sodium via phenylcyclohexenes to phenylcyclohexane. This compound is also generated with magnesium (see Figure 2), however in significantly lower amounts that can hardly account for the significant excess magnesium necessary. Hence, it seems more likely that the excess of this metal needed to accomplish total dechlorinations, either in solid or liquid materials, may be chiefly due to diluting effects of the solid or liquid matrix. Further investigation pertaining to this issue is under way.

Regarding magnesium, experimental data suggest that multiple, coupled Grignard-Zerewitinoff-like reactions may represent one major reaction path here. Interestingly, common laboratory Grignard reactions are always conducted in solution. These procedures usually require very dry solvents, highly purified reagents and a protective gas atmosphere in order to be accomplished successfully. Even traces of proton donors have to be strictly avoided. Apparently, Grignard or Grignard-like reaction without strictly purified reagents, even in the presence of large amounts of a proton donor, can be carried out with ball milling. Moreover, the intermediary Grignard or Grignard-like compound is subsequently protonated at once by the proton donor.

In general, the reactivity of the base metal governs the reaction time. Regarding DMCR treatment of PCB, the dechlorination power of the deployed metal declines in the following series, qualitatively coinciding with a decreasing reduction potential: Na > Mg > Al >> Zn, Fe. Therefore, the type of the metal *and* the hydrogen source were identified as decisive chemical parameters in the DMCR approach as well. Approaches for speeding up dechlorinations of PCBs in natural soils, filter dusts and transformer oils can be permanently tested and scaled up using highly efficient, eccentric vibratory mills such as an "ESM 234" device (SIEBTECHNIK GmbH, Mülheim/Ruhr, Germany). Further information is available on the Internet (TRIBOCHEM, 2003).

Dechlorination of PCP in solid matter. It was verified that other toxic and recalcitrant polychlorinated aromatics can be rapidly and entirely dechlorinated in the same manner as PCBs. According to Figure 9, PCP was completely converted to phenol as the sole product, simply by using appr. 5 % magnesium and 3 % amine regarding the solid matrix (sand) (w/w). The dechlorination proceeded via partly dechlorinated chlorophenols only: as for PCBs or DCB, partly dechlorinated intermediates could be identified by GC analyses using analytical standards of these compounds during time-resolved measurements.

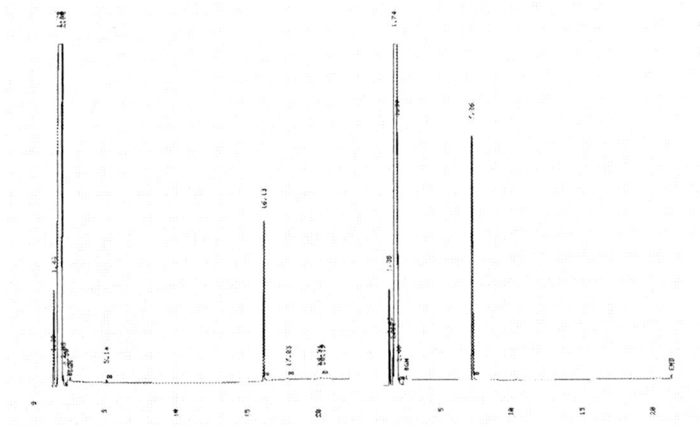

Figure 9: Complete reductive dechlorination of PCP to phenol in sand during milling for *30 min. at room temperature* with magnesium and ethylenediamine: left = prior to treatment in a Retsch S 1000 centrifugal ball mill, right = after treatment (according to GC-FID analysis).

Dechlorination of PCDD/F in solid matter. Treatability studies on dechlorination of PCDD and PCDF in contaminated soils and fly ash showed performance similar to dechlorination of PCBs or PCP in artificial and natural soils (Birke, 2000 and 2001). Besides other metals, sodium and magnesium in combination with different hydrogen donors were tested using centrifugal and vibratory ball mills. Results of one of these studies are outlined in Table 2. Similar destruction efficiencies were obtained after DMCR treatment of soils contaminated by PCDD/Fs, PCBs and polychlorophenols. Apparently, DMCR can render both PCDD/F contaminated fly ashes and soils harmless by efficiently destroying the contaminants in a short time, i.e., as well as already proven for PCBs in natural soils (Birke, 2000 and 2001).

Table 3: DMCR destruction of PCDD/Fs in fly ash (appr. 50,000 ng/kg) deploying sodium or magnesium and different hydrogen donors at room temperature.

	Mg, Ethanolamine	Mg, Amino Acid	Na, Ethylenediamine
PCCD/F Destruction [%]	> 98	98	> 99

4. DISCUSSION AND CONCLUSIONS

Mechanochemistry, the application of mechanical energy to initiate chemical reactions and/or promote and/or alter their course, has been examined during the last 10 years for the mechanochemical dehalogenation of PCBs, PCP, pesticides, and PCDD/Fs in soil, sediments, liquids, and building debris.

Mechanochemical dehalogenation is accomplished by mixing waste material with chemical reagents in a vibratory or ball mills, which are available for throughputs of materials up to several tons per hour. Reactions occur because of the finely divided nature of the reagents and the energy imparted by colliding balls. Observed products and proposed reaction pathways are consistent with radical and organometal intermediates and provide confidence in understanding these reactions. Reaction rates are proportional to the impact energy, which depends on the number of colliding balls, rotational speed, and the ratio of chemical reagents to waste constituents. During the 1990's, Australian scientists patented a ball mill process using calcium oxide or metals as the primary reagents for destroying highly concentrated DDT and PCB-containing wastes (Donecker et al., 1994; Rowlands et al., 1994). Reaction rates are slow, of the order of 6-12 hours, and reaction products from this process (Hall et al., 1996) are complex and present a potential regulatory and public concern.

The DMCR method uses small amounts of metals and hydrogen sources, such as magnesium and alcohols, polyethers or amines, to treat the waste. PCBs in solid or liquid matrices can be destroyed to non-detectable concentrations (GC MS, GC ECD) in a period of minutes to hours. The reaction pathway for this process is well-defined as demonstrated by the isolation and recovery of over 90 % of biphenyl, the dechlorinated product of PCB. The other major reaction product is magnesium chloride. DMCR, which has been carried out in laboratory and pilot studies, yields acceptable, defined reaction product at much faster rates than the calcium oxide process. These advantages lead to greater throughput, improved economics, and regulatory and public acceptability. Critical cost parameters are pollutant concentration, metal and promoter concentration, and water content of the waste. For the case of a small ball mill capable of holding 0.85 tons (2,000-10,000 tons/year) of waste, operating costs (utilities, labor, maintenance, parts, chemicals, and analytical) and capital costs (3-year depreciation) are calculated at $ 140-450/ton for waste containing 1,000-10,000 ppm PCB and water contents of 15 %.

DMCR offers the following benefits:

1. A cost-effective treatment alternative for PCB, dioxins, and pesticides for small or remote sites, for which conventional disposal is unattractive. Even highly concentrated pollutants, e.g., pure PCB oils or pesticides such as hexachlorocyclohexane, can be readily converted to valuable, less toxic chemical compounds in high yields under benign conditions, i.e., biphenyl and trichlorobenzene or cyclohexane, resp., which can be rationally utilized in process streams or marketed.
2. An effective method for treating PCB and dioxin residuals from on-site, indirect thermal desorption processing, liquid residues, or building debris.
3. A treatment method for dioxins in solids, for which no current treatment facility exists.
4. Reduced potential emissions to the environment increase regulatory and public acceptability because the process operates at ambient conditions in a closed system and generates no gaseous emissions.

REFERENCES

Aresta, M., Pastore, T., and DeStefano, L. (2002). Ternary Hydrides for the Direct Dehalogenation of PCBs In Polluted Soil: A Solid State Technology. *International Workshop On "Mechanochemical Processes For the Environment" Science and Industrial Applications, Feb. 15th 2002, Metea Research Centre, Bari, Italy,* Workshop Proceedings.

Aresta M, Dibenedetto A., and Pastore T. (2003). Mechanochemistry: An Old Technology with New Applications to Environmental Issues. Decontamination of Polychlorobiphenyl-contaminated Soil by High Energy Milling in the Solid State with Ternary Hydrides." *Environmental Chemistry Letters*, 2003, in press.

Birke, V. (1996). Zur reduktiven Dehalogenierung aromatischer Halogenverbindungen. *Ph.D. Thesis*. University of Hannover, Germany (in German).

Birke, V. (1998). Dekontamination polyhalogenierter Schadstoffe durch chemische Verfahren. *TerraTech*, Mainz, Germany, 5/1998: 52-57 (in German).

Birke, V. (2000). Economic and Ecologically Favorable Destruction of Polyhalogenated Pollutants Applying the DMCR Technology (DMCR = Dehalogenation By Mechanochemical Reaction), *Contaminated Soil 2000, Proceedings of the Seventh International FZK/TNO Conference on Contaminated Soil*, 18-22 September 2000, Leipzig, Germany. Thomas Telford Publishing, London, ISBN 0 7277 2675 7, 1330-1331.

Birke, V. (2001). Economic and Ecologically Favorable Destruction of Polyhalogenated Pollutants Using the DMCR Technology (DMCR = Dehalogenation By Mechanochemical Reaction). In: Vijgen, J., Pruszynski, S., Stobiecki, S., Sliwinski, W. (Eds.), 6th International HCH and Pesticides Forum, 20-22 March 2001, Poznan, Poland, Forum Book, ISBN 83-913860-7-4, 535-541. <http://www.6thHCHForum.com>

Birke, V. (2002). Reductive Dehalogenation of Recalcitrant Polyhalogenated Pollutants Using Ball Milling. In: A.R. Gavaskar and A.S.C. Chen (Eds.), Remediation of chlorinated and recalcitrant compounds – 2002. Proceedings of the Third International Conference on Remediation of Chlorinated and Recalcitrant Compounds (Monterey, CA, May 2002). ISBN 1-57477-132-9 (computer optical disk), published by Battelle Press, Columbus, OH. <http://www.battelle.org/bclscrpt/bookstore/2002chlorinated.cfm>

Cao, G., Doppiu, S., Monagheddu, M., Orru, R., Sannita, M., Cocco, G. (1999). Thermal and Mechanochemical Self-Propagating Degradation of Chloro-organic Compounds: The Case of Hexachlorobenzene over Calcium Hydride. *Ind. Eng. Chem. Res.*, 38(9): 3218-3224.

Donecker, P., McCormick, P.G., Street, R., and Rowlands, S.A. (1994). Toxic Material Disposal. PCT WO94/14503.

Gock, E., Opel, M., and Mayer, J. (2001). Mechanical Activation of PCDD/PCDF-Containing Material. *Organohalogen Compounds*, 54: 185-188.

Hall, A.K.; Harrowfield, J.M., Hart, R.J., and McCormick, P.G. (1996). Mechanochemical Reaction of DDT with Calcium Oxide. *Environ. Sci. Technol.*, 30: 3401-3407.

Hawari, J. (1992). Regioselectivity of Dechlorination: Reductive Dechlorination of Polychlorobiphenyls by Polymethylhydrosiloxane-Alkali Metal. *J. Organomet. Chem.*, 437: 91-98.

Heinicke, G. (1984). *Tribochemistry*. Akademie-Verlag, Berlin.

Ikoma, T., Zhang, Q., Saito, F., Akiyama, K., Tero-Kubota, S., and Kato, T. (2001). Radicals in the Mechanochemical Dechlorination of Hazardous Organochlorine Compounds Using CaO Nanoparticles. *Bull. Chem. Soc. Jpn.*, 74: 2303-2309.

Janssen, H. and Wille, A. (2002). Untersuchungen zur mechanochemischen Zerstörung von polychlorierten Biphenylen (PCB). *Diploma Thesis*, University of Applied Sciences of North East Lower Saxony, Germany (in German).

Loiselle, S., Branca, M., Mulas, G., and Cocco, G. (1997). Selective Mechanochemical Dehalogenation of Chlorobenzenes over Calcium Hydride. *Environ. Sci. Technol.*, 31: 261-265.

Rowlands, S.A., Hall, A.K., McCormick, P.G., Street, R., Hart, R.J., Ebell, G.F., and Donecker, P. (1994). Destruction of Toxic Materials. *Nature*, 367: 223.

Smith, J.G., and Ho, I. (1972). Reductive Dehalogenations by Alkali Metals and Sodium Naphthalenide. Capture of Solvent-derived Intermediates. *J. Org. Chem.*, 37: 4260-4264.

Smith, J.G., and Ho, I. (1973). Reductive Dehalogenation of Aryl Chlorides by Alkali Metals and Sodium Naphthalenide. Evidence for Radical Intermediates. *J. Org. Chem.*, 38: 3601-3603.

TRIBOCHEM. Web site (2003). <http://www.tribochem.de> (Feb. 1st, 2003).

ECOLOGICAL RISK ASSESSMENT FOR AS-CONTAINING CHEMICAL WARFARE AGENTS – STATUS AND PERSPECTIVES

Tina Vollerthun[1], Wolfgang Spyra[1]
[1]*Brandenburg University of Technology Cottbus, Chair of Chemical Engineering and Hazardous Wastes, Postfach 10 13 44, 03013 Cottbus (Germany)*

Abstract: The Chemical Weapons Convention of 1992 does not regulate contaminations of soil and water by starting, by-products and degradation products of chemical warfare agents that do not pose an acute danger but have harmful long-term effects. The financial means for removing these risks are often missing. A solution for this problem could be the application of adapted microorganisms that are able to biodegrade organoarsenic compounds. Examinations with the model substance triphenylarsine have shown that such a degradation is basically possible.

Key words: arsenic, triphenylarsine, biodegradation, chemical weapons, chemical warfare agents

1. INTRODUCTION

Arsenic is a semi-metal and its most important oxidation numbers are +III and +V. As(III) is much more toxic than As(V). Both species occur in inorganic as well as in organic compounds. Organoarsenic compounds with trivalent arsenic were produced as pesticides and as chemical warfare agents in the first half of the 20th century. There are a number of arsenic containing substances in the chemical weapons' program that are of human- and ecotoxicologic relevance. For a part of these arsenic containing warfare agents, the duty of disposal is regulated in the Convention on the Prohibition of the Development, Production, Stockpiling and Use of Chemical Weapons and Their Destruction of 1992 and its ratification in the year 1997. Completely ignored were here contaminations of water and soil with substances that are not listed. These chemical weapon substances are well-known from battlefields, testing and training areas as well as from production facilities and brownfields. Included are contaminations by starting, side and degradation products of chemical warfare agents and contaminations that need not be removed immediately because there is no acute danger but that pose nevertheless a long-term risk because they can

V.M. Kolodkin and W. Ruck (eds.),
Ecological Risks Associated with the Destruction of Chemical Weapons, 129–133.

enter the human body via the food chain or secondary paths in the environment. The financial means necessary for remediation are often missing. A possible solution is the use of adapted autochthonous microorganisms for biodegradation of organoarsenic compounds. The end-of-pipe products need to be tested with regard to their effects on and risks posed to humans.

2. MATERIALS AND METHODS

The model substance tested for its biodegradability is triphenylarsine. It is not a chemical warfare agent itself but in its chemical and biological behavior similar to chemical warfare agents like phenarsazine chloride (Adamsite, DM), diphenylchloroarsine (Clark I, DA) and diphenylcyanoarsine (Clark II, DC). Triphenylarsine was used in arsine oil, which is a technical mixture of 50 % phenylarsine dichloride, 35 % diphenylchloroarsine, 5 % triphenylarsine and 5 % arsenic trichloride and was used as an additive for bis(2-chloroethyl)sulfide (mustard gas) in order to lower the freezing point [1-3].

Autochthonous microorganisms isolated from soil and water samples from different contaminated sites were used to test the biodegradation of the model substance.

The culture medium used was a mineral salts medium (MSM) of the following composition: $Na_2HPO_4 \cdot 2H_2O$ 2.0 g/l; KH_2PO_4 0.8 g/l; $(NH_4)_2SO_4$ 1.8 g/l; $MgSO_4$ 0.123 g/l.

Chemical analyses were carried out with High Performance Liquid Chromatography (HPLC) using an ODS Hypersil RP 18 column (125 x 3 mm) and a UVD 170 S detector at wavelengths UV 1 = 254 nm and UV 2 = 210 nm. The eluents were A = 95 % water/5 % acetonitrile and B = 95 % acetonitrile/5 % water, the flow rate was 400 $\mu l*min^{-1}$, and the injection volume was 25 μl.

The support material used in the assays with fixed microorganisms was a soft brown coal, an intermediate form of wood and lignite.

3. RESULTS AND DISCUSSION

3.1 Fixed-bed reactor

Five fixed-bed bioreactors were prepared with 10 g of dry soft brown coal each and a total circulating volume of 1 liter of MSM containing 10 ml of a solution of 10 g triphenylarsine per liter ethanol. To each one a different microbial culture was added. The reactors were run at room temperature for 16 weeks.

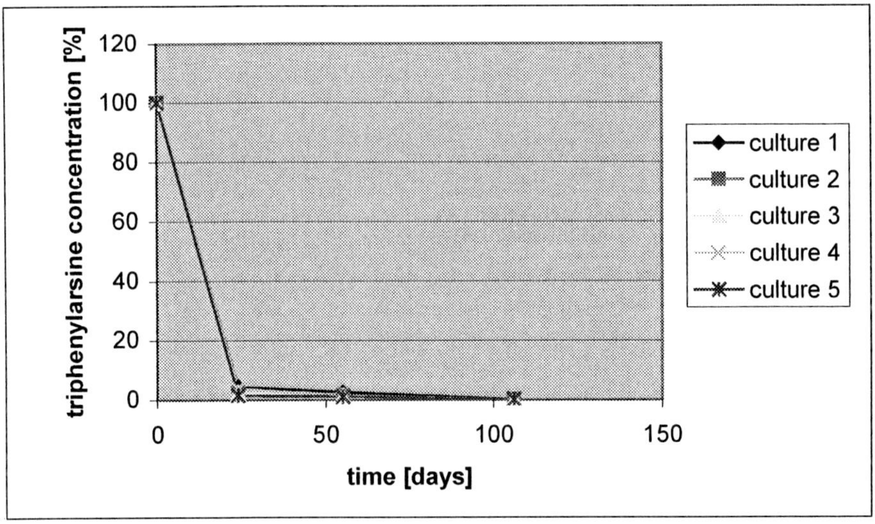

Figure 1: Triphenylarsine concentration in fixed-bed reactors

As can be seen in figure 1, the triphenylarsine concentration in the solution decreases in all cases below the detection limit. To test whether this is partially due to absorption processes, the concentration on the soft brown coal was also determined at the end of the test period. In the case of culture 2, no triphenylarsine could be found; for culture 1 the amount was less than 0.03 % of the original amount; for culture 4 less than 1 %, and for the others slightly more than 2 %. This proves that in all cases an almost complete biodegradation of the model substance has taken place.

3.2 Stirred reactor

A stirred reactor without support material was run in batch mode over a period of one week. It contained 1 liter of MSM with 10 ml of a solution of 10 g triphenylarsine per liter ethanol. Through continuous stirring and inflow of air a saturation of dissolved oxygen was assured. The microorganism culture used in this test was culture 5. Samples were taken every day and analyzed by HPLC. A continuous decrease of the triphenylarsine concentration was observed which reached 0 mg/l after five days.

Figure 2 shows two overlaid HPLC chromatograms where the disappearance of the triphenyl-arsine peak (red) and the occurrence of unidentified metabolite peaks (black) are clearly visible.

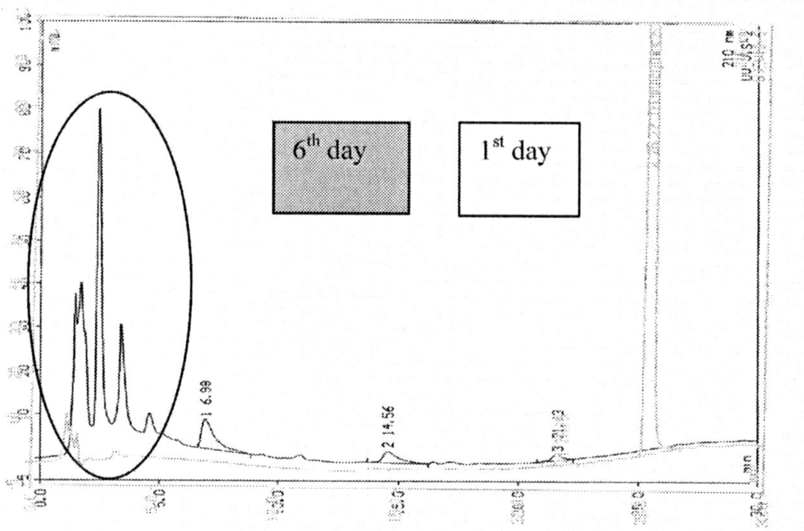

Figure 2: HPLC chromatograms

4. CONCLUSIONS

The results of the executed tests clearly show that a biodegradation of organoarsenic compounds is possible. A number of metabolites are formed , which have not been identified yet. Triphenylarsine oxide and phenylarsonic acid are possible metabolites since they were found by Hofmann et al. [4] in similar examinations. The number of different peaks found in the chromatograms indicates though that there must have been formed other compounds also. In percolation experiments with contaminated soil Köhler et al. [5] observed a microbially induced release of a number of organoarsenic compounds.

Up until now, it has not been possible to completely mineralize organoarsenic compounds but further investigations are being made to remedy this.

References

[1] Haas, R. (1996). Blaukreuzkampfstoffe – Chemisches Verhalten und humantoxikologische Bedeutung von Diphenylarsinverbindungen. Teil I: Chemisches Verhalten. Umweltmedizin in Forschung und Praxis 1(4):183-189

[2] Haas, R., Schmidt, T.C., Steinbach, K., v. Löw, E. (1998). Chromatographic deter-mination of phenylarsenic compounds. Fresenius Journal of Analytical Chemistry 361:313-318

[3] Martinetz, D. (1993). Arsenorganische Verbindungen in Rüstungsaltlasten. TerraTech 4:36-40

[4] Hofmann, K., Hammer, E., Köhler, M., Brüser, V. (2001). Oxidation of triphenylarsine to triphenylarsineoxide by *Trichoderma harzianum* and other fungi. Chemosphere 44:697-700

[5] Köhler M., Hofmann, K., Völsgen, F., Thurow, K., Koch, A. (2001). Bacterial release of arsenic ions and organoarsenic compounds from soil contaminated by chemical warfare agents. Chemosphere 42:425-429

3

SESSION II: MONITORING

PERSONAL MONITORING OF OLD CHEMICAL WEAPONS' DISMANTLING OPERATIONS AT POELKAPELLE (BELGIUM)

Herbert C. De Bisschop[1], Christel Meert[1], Bart R. Smedts[1], Christiaan P. Perneel[2]

Royal Military Academy, [1]Department of Chemistry, [2]Department of Mathematics, 30 Avenue Renaissance, 1000 BRUSSELS (Belgium)

Abstract: This article focuses on safety aspects of the dismantling of Old Chemical Weapons in Belgium, in particular of the shell type 77 mm filled with mustard gas (HD) and frequently chlorobenzene. Experiments were set up in order to determine whether the exposure limits for these chemicals (long-term as well as short-term) were exceeded during the dismantling process. In this process, two distinct operations were identified: (i) task 1 in which drilling and sampling are the main operations; (ii) task 2, consisting of a milling operation in order to separate the head from the shell body, followed by emptying and packing. For both subtasks the measured values are in agreement with a lognormal distribution. Significantly higher values for HD and for chlorobenzene concentrations are associated with task 2. A long-term respiratory overexposure might be excluded for chlorobenzene; for HD, however, a respiratory protection factor of at least 200 is indicated. To evaluate possible short-term overexposure via the skin, C.t values were measured for a sequence of task 2 operations. The results cannot be compared with a specific statistical distribution due to the fact that (i) these values are not truly independent, and (ii) the number of measurements is too limited to apply other inferential statistical methods. From the raw data, however, one can conclude that the upper confidence limit value of the 95th percentile remains lower than the accepted limit of 5 $mg.min/m^3$.

Key words: sulphur mustard, chlorobenzene, occupational exposure, old chemical weapons, dismantling

1. INTRODUCTION

The Chemical Weapons Convention defines Old Chemical Weapons (OCW) as chemical weapons that have been produced before January 1, 1925 or as chemical weapons produced between 1925 and 1946 that have deteriorated to such an extend that they can no longer be used as chemical weapons; under this Convention OCW can be treated as toxic waste [1].

V.M. Kolodkin and W. Ruck (eds.),
Ecological Risks Associated with the Destruction of Chemical Weapons, 137–152.

However, the destruction of such weapons is not obvious: the degree of corrosion and the possibility that the explosive train is still active make automated reverse assembly impossible. The choice of a safe disposal method is always a compromise between realistic and scientifically justified levels of emission and exposure on the one hand, and the hazard of unnecessary delays in the destruction of these weapons on the other hand [2]. In Germany [3], the United Kingdom [2] and Belgium, where OCW from the First World War are still retrieved on a regular basis, dismantling has been chosen as a basic step in the process of their destruction.

In Belgium, OCW are unearthed together with, and indistinguishable from, conventional munitions. Shells classified as suspected toxic are confirmed as being toxic by the non-destructive technique, x-ray analysis. Neutron Activation analysis provides some additional information on the elemental composition of the toxic payload. As a first step in their dismantling, the toxic contents of the shells are completely identified by sampling through a hole drilled in the munition, followed by chemical analysis. The head of the shell, containing the fuse and burster charge, is separated from the shell by milling, and the agent can be drained or otherwise removed. Both drilling and milling are remote-controlled operations carried out in specific bunker structures, are kept under a negative pressure of 50 Pa. During remote-controlled operations workers are staying in a waiting zone. However, sampling, separation of fuse and burster charge from the chemical payload, draining and packing of the remainder of the shell are manual operations, where operators are potentially exposed to the toxic contents, requiring appropriate protection measures for the operators.

An important part of the shells that are to be dismantled contains sulphur mustard [(bis)-2-chloroethylsulphide, code HD], frequently in combination with chlorobenzene. Sulphur mustard is an alkylating agent that was used intensively as a vesicant towards the end of the First World War. Its acute and chronic toxicity effects have been reviewed [4]. For risk evaluation two exposure limits are used. For short-term exposure of skin to vapour, a minimum effective dose of 50 mg.min/m^3 has been reported [5]. To take into account enhanced effects at higher temperatures and higher humidity [4], a short-term exposure limit (STEL) value, expressed as concentration x time (C.t) of 5 mg.min/m^3 has been tentatively accepted. Cancer is seen as the most problematic long-term effect of inhalation exposure to HD vapour. Via low-dose extrapolation of animal data it can be calculated that an exposure during a working life (40 years) at 0.2 μg/m^3 could be associated with 4 extra cancers in 100,000 workers [6]. Chlorobenzene is a solvent that was added to lower the freezing point of HD. Its limit value expressed as TWA over an 8h working day is 47 mg/m^3 [7].

In this contribution we report on the exposure of workers in the dismantling facility for OCW at Poelkapelle (Belgium). One of the basic questions to be answered before the facility was allowed to start up, was whether the use of the NBC suit, supplemented with a positive-pressure breathing apparatus, kept the risk for long-term effects at an acceptable level. For unprotected operators exposure is estimated by using active personal sampling. The exposure for protected operators is then assessed by taking into account the protection factors of the protective equipment and compared with an exposure level at which no toxic exposure is likely to occur or at an accepted risk level [8]. The monitoring for long-term exposure covered roughly one year, during which a variety of shells of various types were dismantled. Short-term exposure was evaluated at later stage by monitoring the part of the dismantling process that leads to the highest concentrations, i.e. the activities associated with the milling operation.

2. OUTLINE OF DISMANTLING OPERATIONS

2.1 Description of the toxic shell type involved

The shell type treated in this contribution is a 77 mm LFKGr (Lange Field Kanone Granate) "Yellow Cross", i.e. filled with mustard gas and, predominantly, chlorobenzene added to lower the freezing point. Figure 1 shows a section of the shell. Table 1 lists its main characteristics.

2.2 Dismantling

At present, non-destructive testing does not allow to fully identify the contents of a chemical shell. Therefore, a first set of operations (task 1) in the dismantling process is drilling and sampling (5 to 10 mL with a glass pipette), followed by filling the hole with a plug and storing the shell in a refrigerator until the next set of operations (task 2) consisting of a milling operation to separate the head (with fuse and burster charge) from the shell body, well above the liquid level. Fuse and burster charge are soaked in *n*-butanol for at least 24 h. and stored for disposal by covered-earth detonation. The chemical content of the body is transferred to a 30 L high-density polyethylene (HDPE) container, the shell body is packed and both parts are stored awaiting their destruction at a toxic waste incinerator in accordance with European standards [9].

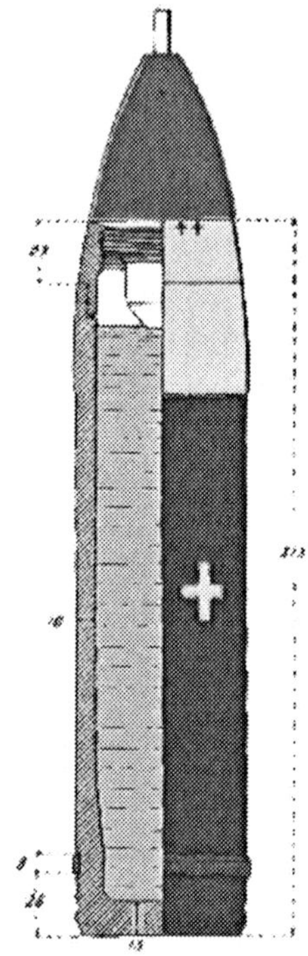

Figure 1: Cross section of a 77 mm LFK Gr (Lange Field Kanone Granate) GelbKreuz ("Yellow Cross").

Table 1: Main characteristics of 77 mm LFK Gr

Calibre (mm)	*77*
Length, without fuse (mm)	*313*
Wall thickness (mm)	*10*
Overall weight (kg)	*6.875-7.330*
Toxic payload volume (cm^3)	*670*
Explosive charge (kg)	*0.020*

2.3 Protective measures

The rooms where chemical agents might be present stay under negative pressure and are continuously ventilated. For the last set of experiments (monitoring of selected activities), local exhaust ventilation was used during milling and emptying of the shell body (flow rate is 40 m^3/h).

Throughout the whole dismantling operations, workers are dressed in protective suits. Two types of suits are used: (1) a fully encapsulating suit comprising of active carbon coated underwear (Blücher type Saratoga) and an impermeable overgarment containing 14 layers of laminate [10] and (2) a four layer semi-permeable NBC protective suit combined with the underwear cited under (1). Both suits are don with rubber boots and butyl rubber gloves and used with a closed circuit (positive pressure) breathing apparatus with a compressed-air canister carried on the operator's back. For the Chemrel Max GT garment (suit (1)) a 2 h limit is set including the time for donning and doffing.

Under normal operating conditions operators wear the NBC suit, except when an incident has occurred or when the automated HD alarm indicates concentrations above 200 mg/m^3. In that case, colleagues dressed in the impermeable suit type replace operators in NBC suits.

3. EXPERIMENTAL SET-UP

3.1 Monitoring over a complete working day

A team of two operators performs the tasks mentioned under 2.2 above that contain either manual handling or automated processing, or both. In this set of measurements a portable sampling system was passed on from one team to the next until the end of a working day as a means of simulating the exposure of one (hypothetical) team that would have completed the whole activity on its own. This is, however, a worst case that is impossible under the current rules of procedure: an operator works an average of one hour per day in NBC protective clothing. The tasks performed over a working day are counted as tasks 1 and 2, referred to above.

3.2 Monitoring for short-term exposure

Monitoring over a complete working day results in a time-weighted average and it gives no information on peak concentrations that may occur during short periods of time and may result in doses leading to acute effects. From the results obtained under 3.1, it can be concluded that task 2, which includes the emptying of the shell, will lead to the largest peak concentrations. Task 2 can be seen as being divided in two sub-tasks: task 2A consists of bringing the shell to the milling apparatus, clamping and securing it; task 2B is done after remotely controlled milling and consists of taking the head of the shell off and bringing it to a decontamination bath, unclamping the shell body, emptying it in a HDPE container, closing off the upper shell body with a towel and storing it in another HDPE container.

For each shell dismantled, the main operator is equipped with active personal air sampling equipment that is changed after each subtask. Moreover, one of the operators was also equipped with passive sampling as indicated under 4.1.3. From the analytical results, a time-weighted average exposure is calculated for each shell dismantled.

4. ANALYTICAL CHEMISTRY

4.1 Samples

4.1.1 Active Personal Sampling

Active sampling is done with TENAX-GR tubes, ¼'' external diameter, 3.5'' length (Supelco 25054) and a GILIAN LFS 113 DC in constant pressure mode. Typical flow is between 100 and 200 mL/min. Except for the sampling over a whole day, each sampling was done in duplicate.

4.1.2 Active Area Sampling

Active sampling in specified areas is done over a complete working day with identical equipment as for active personal sampling. Flow is set at 70-100 mL/min.

4.1.3 Passive Personal Sampling

Passive sampling is done with PADS (Passive ADsorption System-DEVCON 037-002101-115 Natick Sampler Unfilled). These are available with standard dimensions of 17 x 26 mm but are to be filled with the appropriate adsorbent. The PADS are prepared as follows:

a) The adsorbent for the PADS is preconditioned as follows: 300 mg of TENAX-GR polymer (60/80 mesh (Supelco 25054) is put in an empty Perkin Elmer ATD tube (Sigma Aldrich 25049) and sealed with a temporary plug and desorbed in a Perkin Elmer ATD 400 at 320°C, under a Helium flow of 50 mL/min for 30 min. The preconditioned material is kept in a closed container (moist free) and in the refrigerator at -20°C.

b) PADS are numbered, weighed empty and labelled.
c) PADS are filled with 40-50 mg of preconditioned TENAX and weighed again. The amount of added adsorbent is calculated.
d) The filled PADS are sealed with tape and covered with a TEFLON cover window (10x20 mm) to reduce the error in exposed surface.

Sampling procedure of the PADS is as follows:

a) PADS are attached to personnel at the outside as well as under the NBC-gear in triplicate. The PADS under the gear are fixed with hypoallergenic tape directly on the skin in the neck/facial area under the face piece of the mask (Figure 2).
b) PADS are combined per cluster of three to increase the total surface exposed. One cluster is placed in the neck of the personnel and a second one is placed on the front part of the NBC-gear.
c) After sampling, the PADS are immediately removed and put in an active carbon envelope. The activity completed, the contents of a cluster is transferred in an empty Perkin Elmer ATD tube (Supelco 25054).
d) The filled ATD tubes are cooled, transported to the laboratory for analysis. The samples that cannot be analysed immediately are stored in the refrigerator at –20°C.

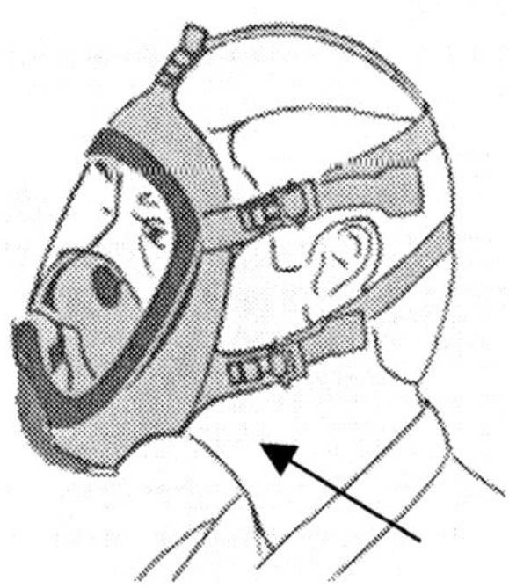

Figure 2: The arrow indicates where the PADS are placed in the neck area.

4.2 Analysis

4.2.1 Analysis of sampled TENAX adsorbent

After sampling, the tubes are unsealed and spiked with internal standard (1 µL of a 4-bromo fluorobenzene solution in dichloromethane, 15.94 ng/µL). Analysis is performed by thermal desorption/cold trapping followed by gas chromatographic separation and mass-spectrometric detection. Tubes are thermally desorbed for 10 min at 300°C under a Helium flow of 3 – 20 ml/min, cold trapping at –30°C, followed by flash heating 40°C/s to inject the sample on the column (DB-5MS (95 % methyl, 5 % phenylpolysiloxane) 30 m x 0,32 mm ID x 0,5 µm df. Column flow is Helium 2.8 mL/min, temperature program is 35°C (5 min), 20°C/min to 180°C and 60°C/min to 280°C. Blank tubes (containing only adsorbent) and tubes spiked with known amounts of HD and chlorobenzene are added for control and calibration purposes.

Single Ion Monitoring (SIM) is done for chlorobenzene (m/z 112), HD (m/z 109 or 158) and 1-bromo-4-fluorobenzene (m/z 175).

5. RESULTS

5.1 Detection limits and calibration

The reliable limits of detection of the agents for the whole analytical procedure (DLOP) are determined according to Burright *et al.* [12] from a calibration in the low range by using the formula:

$$\mathrm{DLOP} = \frac{3 \cdot S_{Y\text{-}X}}{A}$$

where $S_{Y\text{-}X}$ is the standard error of the estimate for the regression line
A is the analytical sensitivity (slope of the regression line)

A typical low-range calibration for the purpose of determining the DLOP is shown on figure 3.

Values obtained are absolute quantities, but they can also be expressed as concentrations by taking into account a typical sampling flow rate of 100 ml/min and a sampling time of 30 minutes (table 2). The lower values for chlorobenzene are mainly due to its better chromatographic peak shape. Obviously, the value obtained for sulphur mustard at m/z = 158 is lower than that for the base peak at m/z = 109. However, quantification at the base peak is not possible for the analysis of PADS fixed onto the skin, due to background interference.

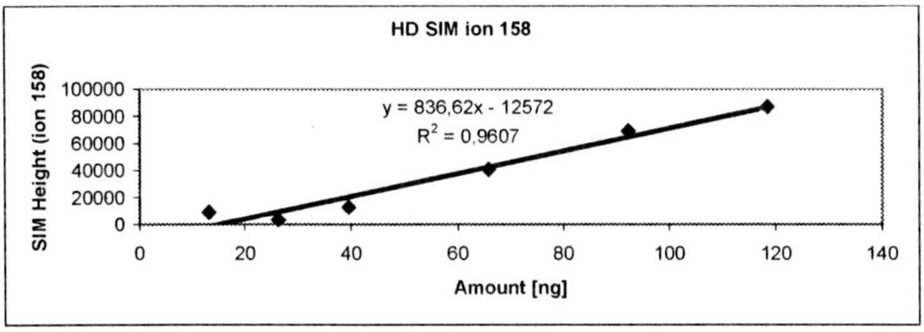

Figure 3: HD calibration for m/z 158, height of SIM peak versus amount of HD. TENAX tubes are spiked with known amounts of HD and analysed as described in the text.

Compound	selected ion	DLOP	
		(ng)	($\mu g/m^3$)
HD	109	9	3
HD	158	19	6
Chlorobenzene	112	0.12	0.04

Table 2: Reliable detection limits of the analytical procedure (DLOP) for HD and chlorobenzene.

5.2 Exposure data

5.2.1 Monitoring over a complete working day

The results, expressed as TWA concentrations, cover about one year of operation and are restricted to those days when 77-mm "Yellow Cross" shells were dismantled, although other shell types may also have been dismantled on those same days. The values lie between "not-quantifiable" and 210 $\mu g/m^3$ for HD and 2752 $\mu g/m^3$ for chlorobenzene. The higher maximum value of chlorobenzene is explained by its lower boiling point. From the results, two sub-groups are constructed, one for the concentrations obtained on those days when only task 1 was performed, and another one related to those days when only task two was done. Both subgroups are in agreement with a lognormal distribution, as illustrated by Q-Q-plots (figure 4). The calculated lognormal distribution of both subgroups is given in Figure 5.

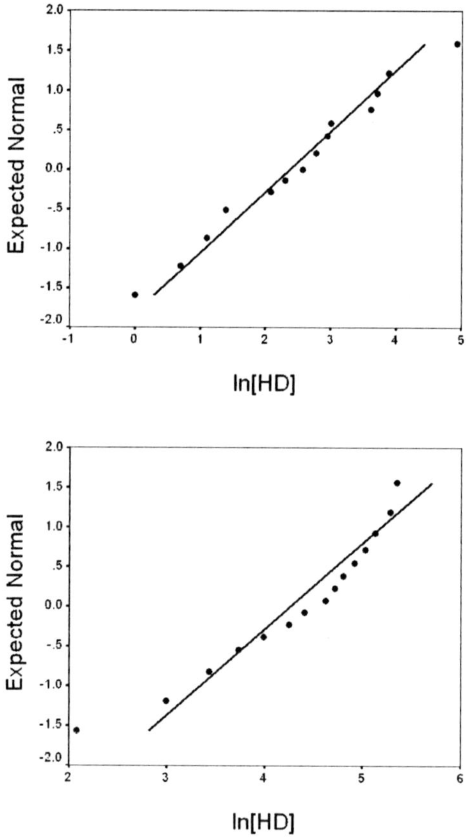

Figure 4: Q-Q plot of the log-transformed data of [HD] for task 1 (left) and task 2. The observations lie approximately on the reference line. The normality of the log-transformed data is confirmed by the Kolmogorov-Smirnov test ($p > 0.2$ for both tests) and the Shapiro-Wilk test ($p=0.563$ for task 1 and $p=0.242$ for task 2)

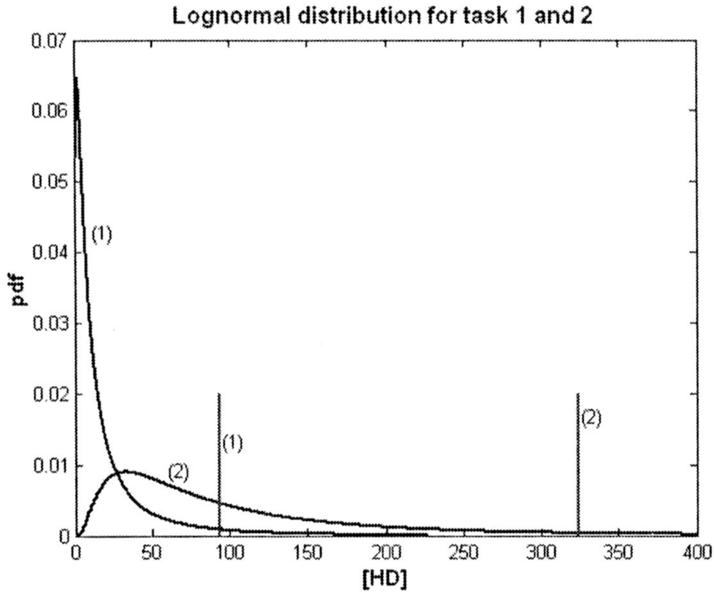

Figure 5: Estimated lognormal distribution (probability density function, pdf) of tasks 1 and 2. The 95th percentile values are indicated by the vertical lines. Numbers in brackets identify the task numbers. [HD] is the concentration of sulphur mustard, expressed in $\mu g/m^3$.

The central position of the concentrations of HD and chlorobenzene is significantly higher for task 2 than for task 1 (Mann-Whitney test). The upper confidence limit of thc 95th percentile is usually taken as the reference to be compared with the log-term exposure standard. For chlorobenzene, one may accept that there is no overexposure at the 95th percentile. For HD, the upper confidence limit of 627 $\mu g/m^3$ has to be compared [13] with the standard of 3 $\mu g/m^3$, taking into account the protection factor of the positively pressurised mask.

Table 3: Statistical values of the lognormal distributions of the concentrations of HD and chlorobenzene for tasks 1 and 2. All concentrations are in $\mu g/m^3$

Agent	HD		Chlorobenzene	
	median	95th percentile (in brackets, thc 95% confidence limits)	median	95th percentile (in brackets, the 95% confidence limits)
task 1	11.0	94 [36;240]	30.0	804 [178;3561]
task 2	73.7	324 [161;627]	221.4	1147 [539;2386]

5.2.2 Monitoring for short-term skin exposure

For the highest concentrations, obviously associated with task 2, it is worthwhile to investigate whether the combination of an insufficiently tight fitting at the hood cover of the overgarnment and the "pumping" effect of the clothing by the movements of body and arms may lead to skin exposures in the neck area exceeding the STEL threshold. Therefore, as a second set of experiments, task 2 is monitored for the purpose of measuring short-term time integrated (C.t) –values. Here again, the operator, carrying out the manual handling, is equipped with active personal air sampling equipment that is changed after each subtask. Moreover, one of the operators is also equipped with passive sampling, done with PADS as described earlier. PADS are attached to the outside as well as under the NBC-gear. By comparing the quantities found at the outside of the overgarnment with those determined at the inside, protection factors for the neck frontal area can be calculated. Protection factors are expected to be dependent on the type and intensity of activities and on the strictness with which the donning procedure is followed. Values are found to be 10^2 , with one low value of 4.

Table 4 : Summary descriptive statistics of the C.t values obtained for HD during task 2. All units (except for N, number of samples) are in µg.min/m^3.

Mean value	Standard Deviation	Minimum value	Maximum value	95th percentile	N
1443.4	1062.3	35.6	3877.5	3760	22

It is impossible to statistically compare the empirical results with a basic statistical distribution, due to the fact that concentrations, generated by the operations associated with the milling of a shell (spills, splashes, etc.) may influence the next measurements. In that respect, the values, while reflecting the real situation, may not be truly independent. On the other hand, the number of measurements is too limited (11 measurements per day during two non-consecutive days) to allow for considering other inferential statistical methods. Summary descriptive statistics of the raw data are given in table 4. It is seen that, for dismantling the 77 mm "Yellow Cross" shells, the 95th percentile of C.t measurements (3.76 mg.min/m^3) is still lower than the accepted 5 mg.min/m^3, not even taking into account any protection factor.

Table 5: Average concentrations obtained from static area measurements during milling operations (11 shells per series) performed to monitoring for short-term exposure. Time-weighted average values of personal air samples calculated for all operators involved over the whole time span of the series. Concentrations are given in μg/m³.

series	HD		
	TWA	at milling equipment	at exhaust
1	126	337	115
2	399	514	394

For the whole duration of the short-term exposure measurements, portable sampling equipment was also installed in the milling room at the level of the milling blade and at the level of the exhaust ventilation of the milling room. Table 5 shows the results compared with TWA values calculated for the manual operations involved in the milling/packing operations. It is seen that, for each series, TWA values found for HD are close to the values measured at the exhaust ventilation, but lower than the values obtained at the milling equipment. This result can be explained by assuming a systematic contamination (spill) due to the milling operation, generating a continuous vapour concentration that is gradually diminishing towards the exhaust.

6. CONCLUSIONS

The conclusions only apply to the dismantling of 77 mm "Yellow Cross" shells under the procedures and collective protection measures that were in use at the time of the experiments.

Manual operations associated with drilling result in statistically less exposure than tasks associated with shell milling.

Long-term respiratory exposure is evaluated from the results of active sampling over a full working day of dismantling 77 mm OCW. To protect the 95th percentile from long term adverse effects of sulphur mustard, a respiratory protection factor of at least 200 is needed.

Short-term skin exposure to sulphur mustard is estimated from consecutive milling of 11 shells (about half a working day). It was not possible to describe the distribution of the experimentally obtained TWA-values by a statistical law. From the raw data, however, it can be inferred that the UCL value of the 95th percentile remains lower than the accepted limit of 5 $mg.min/m^3$.

ACKNOWLEDGEMENTS

The authors thank J.L. Willems, H.A.F. Veulemans and C. Carton for helpful suggestions and C. Dupont for valuable aid in the analysis. The assistance of E. Veys in the experimental set-up and of V. Matten and C. Patfoort in sampling is highly appreciated.

References

[1] Convention on the Prohibition of the Development, Production, Stockpiling and Use of Chemical Weapons and on their Destruction. Paris (1993).

[2] Manley R.G. (1997), Chemical weapon agent and historic chemical munition disposal: the British experience, in: "The Challenge of Old Chemical Munitions and Toxic Armament Wastes", Kh Lohs and T. Stock, Eds., Oxford University Press, New York, pp.231-240.

[3] Martens, H. (1997), The German programme for the disposal of old chemical weapons, in: "The Challenge of Old Chemical Munitions and Toxic Armament Wastes", Kh. Lohs and T. Stock, Eds., Oxford University Press, New York, pp.166-175.

[4] Watson A.P., Griffin G.D. (1992), Toxicity of vesicant agents scheduled for destruction by the chemical stockpile disposal program, Environmental Health Perspectives, 98: 259-280.

[5] Derivation of health-based environmental screening levels for chemical warfare agents. A technical evaluation. US Army center for health promotion and preventive medicine APG & Toxicology and risk analysis section, Life Sciences Division, Oak Ridge National Laboratory. December 1998.

[6] Willems, J.L. (1999), Report to the Steering Committee on the Destruction of Old Chemical Weapons.

[7] Royal Decree Nr.2002013158 (2002), Belgium Federal Gazette (Moniteur-Staatsblad), 25 October.

[8] Herber, R.F.M., Duffus, J.H., Christensen, J.M., Olsen, E., Park, M.V. (2001), Risk assessment for occupational exposure to chemicals. A review of current methodology, Pure Appl. Chem., 73:993-1031

[9] European Community Directive. Directive 2000/76/EC of the European Parliament and of the Council of 4 December 2000 on the Incineration of waste, Official Journal, L 332/91, 28 December 2000.

[10] Chemrel Max GT, Chemron Inc., USA

[11] Liteanu, C., Rica, I. (1980), Statistical Theory and Methodology of Trace Analysis. Ellis Horwood Limited, Chichester, pp.257-275.

[12] Burright, D., Chan, Y., Eide, M., Elskamp, C., Hendricks, W., Rose, M.C., (1999), Evaluation guidelines for air sampling methods utilizing chromatographic analysis,

OSHA Salt Lake Technical Centre, U.S. Department of Labor, Salt Lake City UT 84115-1802, pp.3-5.

[13] Leidel, N.A., Busch, K.A., Lynch, J.R. (1977), Occupational Exposure Sampling Strategy Manual, US Department of Health, Education and Welfare, Cincinnati (Ohio), pp. 47-49.

RADIO-ECOLOGICAL MONITORING OF THE ENVIRONMENT

Irina Fedotova[1]
[1] *Udmurtia State University, Izhevsk (Russia)*

Abstract: The problem of radio-ecological monitoring of the environment has become particularly topical lately. This is due to the intensive development of the radio-chemical technologies, processing and burial of the waste of nuclear explosions for military and peaceful purposes and accident situations. Various types of radiation can strengthen non-radiation factors of danger. The radio-ecological situation on the territory of the Udmurt Republic has not been studied deeply enough. The absence of the necessary information increases the probability of the 'psychogenic effect', the negative attitude to the construction of facilities for disposal of various types of waste and, in particular, to the facilities for chemical weapon destruction. This contribution shows the necessity of organization on the territory of the Udmurt Republic of not only a system of chemical monitoring of the environment, but also of a complex of radio-ecological investigations, training courses for the population, and provision for access to information on the chemical and radiation situation to the public.

Key words: radio-ecological monitoring, factor of danger, radiation background, psychogenic effect, complex of investigations

Nowadays powerful sources of ionizing radiation are intensively used not only in radiation chemistry, but also in agriculture, medicine, the food industry, and for modifying building materials and many other purposes. For example, sources of ionizing radiation are used in agriculture for improving seed germination by exposure to radiation before sowing, against harmful insects, and obtaining new varieties of agricultural plants. Medicine uses radiation for sterilizing medical supplies and pharmaceutical preparations. Processing wood by ionizing radiation gives wood-plastic materials possessing improved mechanical properties, and fire- and moisture-resistance. The range of applications of radiation plants has extended and the assortment of sources of ionizing radiation has increased, respectively.

In addition, it is known that the Northern-Western area of the Northern Urals was exposed to the radiation precipitation in 1986 as a result of the passage of the northern branch of the Eastern Chernobyl trace [1]. Udmurtia has done no investigations in connection with the passage of the Eastern Chernobyl trace over its territory.

V.M. Kolodkin and W. Ruck (eds.),
Ecological Risks Associated with the Destruction of Chemical Weapons, 153–156.

The intensive development of the radiation-chemical technologies, waste processing and burial, and nuclear explosions for military and peaceful purposes contribute to the biosphere pollution with radiation products. This being the case, the fact that various non-radiation factors of danger can become enhanced under the influence of the powerful radiation fields must be taken into consideration. Radiation combined with the effect of chemical carcinogens leads to a more frequent and accelerated formation of tumors. According to some data low doses of gamma-radiation and chemical carcinogens that do not produce tumors when used separately become carcinogens if used together [2].

The radio-ecological situation on the territory of Udmurtia has not been studied enough and the population has not enough information on the level of the background radiation. This is why the problems of carrying out investigations on the radio-ecological monitoring of the environment and studying the effects of the combined factors of danger of the radiation and non-radiation nature on humans, with attention to the unfavorable effect of the 'psychogenic factor,' are of interest. The probability of the 'psychogenic factor' and the negative attitude towards the construction of the facilities for disposal of various types of waste and chemical weapons increases with the absence of the information on various combined effects from chemical and radiation factors of danger to humans, to which the population can be exposed.

To make a complete analysis of the environmental situation in the Udmurtia area, it is necessary to carry out not only the chemical monitoring of the environment, but also a complex of radio-ecological investigations, to organize training courses for the population, and provide access to information on the chemical and radiation situation.

The radio-ecological investigations should comprise the following:

- monitoring of natural (isotopes Ra, Rn, U, Th, ^{210}Pb, ^{210}Po) and technogenic (^{3}H, ^{90}Sr, ^{241}Pu) radionuclides at the background levels in the receptors of the environment;
- express-analyses of discharges and disposals of non-nuclear enterprises: oil deposits and gas fields, chemical facilities, facilities for disposal of waste, metallurgical plants and scrap, and heat and power plants;
- radiation tests of the sources of potable water and food, for the content of alpha- and beta-radiators (^{90}Sr, ^{226}Ra, ^{228}Ra, ^{210}Pb, ^{210}Po, ^{234}U, ^{238}U);
- precision analysis for the content of radon and thoron in the room air.

To inform the public, it is necessary to create a special site where it would be possible to follow the change in the environmental situation, concentrations of radionuclides in various receptors, and the results of checks of food, and building and other materials for quality.

References

[1] Nifontov, M.G. (2003). Modern Levels of Content of ^{90}Sr and ^{137}Cs in Moss-Lichen Cover of Foothills and Mountain Landscape of Northern Ural. Ekologija 1: 51-55.
[2] Larichev, A.V., Chistov, E.D. (1981). Safety in Radiation Technology. Energoizdat, M.

ION MOBILITY SPECTROMETRY FOR MONITORING THE DESTRUCTION OF CHEMICAL WARFARE AGENTS

Herbert H. Hill[1], Wes E. Steiner[1]
[1]*Department of Chemistry, Washington State University Pullman, Washington 99164-4630 (U.S.A.)*

Key words: Point Detectors, Ion Mobility Spectrometry, Chemical Warfare Agents, Mass Spectrometry, Gas Chromatography

1. INTRODUCTION

Monitoring chemical agents to verify their destruction is not a trivial matter. Although the Chemical Weapons Convention (CWC) that entered into force in 1997 stated that the "*The inspection team shall have the right to perform on-site analysis of samples using approved equipment brought by it*" [2] few if any analyses are made during inspections [1]. The difficulty in making field measurements of CWAs results from requirement of using mass spectrometers. While mass spectrometers provide high specificity to insure the identification of the CWA, they are costly and difficult to operate under field conditions. Alternatively, ion mobility spectrometry (IMS) has been used for field analysis but this method is not sufficiently selective to eliminate numerous false positive responses. The combination of IMS with MS however offers and attractive alternative for the verification of CWA and their degradation products in the field. The addiction of an IMS as the interface to a mass spectrometer reduces the operational requirements on the mass spectrometer while the addition of a mass spectrometer after separation by ion mobility improves the selectivity of IMS.

The chemical threat agents which need to be monitored have a wide range of structural complexity. The simplest of these are the chemical agents such as nerve, vesicant, and blood agents along with their corresponding degradation products. These traditional CWAs having a low molecular weight provide an acute toxicity on the level of 10^{-3} games/person. In addition to traditional CWAs, emerging chemical agents such as toxic industrial chemicals and aerosols must also be detected and identified. More chronic but also more potent chemical agents such as bioregulators and toxins must also be detected and monitored. Bioregulators, such as

V.M. Kolodkin and W. Ruck (eds.),
Ecological Risks Associated with the Destruction of Chemical Weapons, 157–166.

neuropeptides and psychoactive compounds, and toxins, such as cytotoxins and neurotoxins, are non-volatile high molecular weight and consequently cannot be monitored by existing methods developed for the traditional, more volatile CWAs.

In addition to the broad spectrum (15 to 200,000 Da threat spectrum) of compounds which must be monitored for the verification of CWAs; analytical methods are asked to perform these analyses in less than 5 min, provide high sensitivity and reproducibility to reduce the potential for **false negative responses**, exhibit high resolving power to reduce the potential for **false positive responses**, and require little or no sample preparation with low reagent consumption in order to be field deployable. Thus the ideal analytical sensor to take to the field would be small, light weight, require low power, be inexpensive yet provide high selectivity and sensitivity. While this "tricorder" like instrument is not currently available there are a number of analytical instruments that do exists which provide unique contributions to the detection challenge.

2. CURRENT ANALYTICAL METHODS FOR CWA DETECTION

The lightest and most field deployable instruments are "zero dimensional sensors". That is, analytical instrument in which the components of the sample do not need to be physically separated prior to detection. Methods, which fall into this category, include surface acoustic wave spectrometry, the electronic nose, electrochemical methods and immunoassays. Without pre-separation of the sample matrix, however, these methods often become contaminated with environmental interferences, leading to numerous false positive responses or non-reproducible operation.

While more complicated to operate, "one dimensional sensors" provide a physical separation step prior to detection, reducing the potential for environmental interferences and increasing the reliability of the response. One dimensional sensors include gas chromatography, liquid chromatographs, capillary electrophoresis, mass spectrometry and ion mobility spectrometry. For volatile compounds gas chromatography with selective detectors are often used for the detection of CWAs. The availability of phosphorus, nitrogen, halogen and sulfur selective detectors make GC methods especially sensitive and selective for selected CWAs. Liquid chromatography and capillary electrophoresis are used less frequently for non-volatile compounds and are still under development as routine CWA detection methods. Mass spectrometry is often thought of as the ultimate in detection capability; with its high resolution and sensitive detection. Mass spectrometry potentially provides universal detection method. However, the requirement of vacuum operation of mass spectrometers reduces the utility of this method for field operation. Dirty environmental samples can rapidly foul the spectrometer, requiring excessive maintenance. In addition, complex samples lead to selective ionization processes, complex fragmentation patterns, and a multiplicity of mass peaks; which complicates and reduces the reliability of mass spectral data. Ion mobility spectrometry on the other hand, has often been used as a substitute for mass spectrometry for field analysis. Recent designs operating at atmospheric pressure are less susceptible to environmental interference and experience less down time than mass spectrometers. Currently, commercially available instruments exhibit low resolution, leading to the potential of false positive responses, but the higher resolution instruments which have recently been developed offer considerable promised for field application.

3. ION MOBILITY SPECTROMETRY

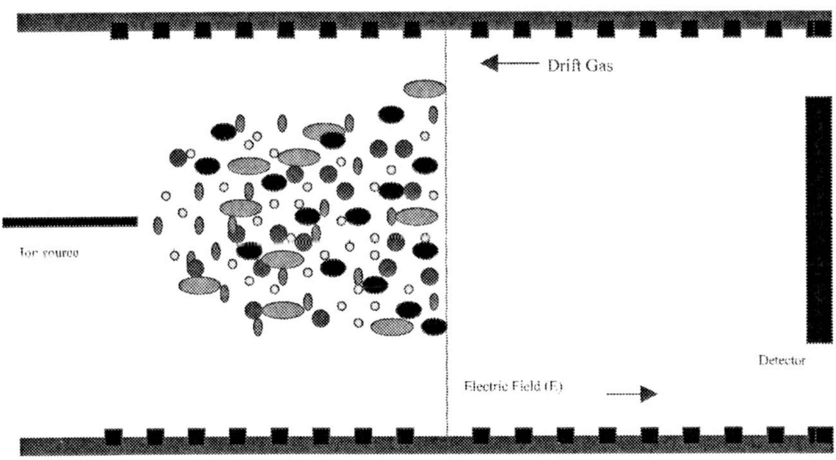

Figure 1: Ionization

Figure 1 shows a schematic diagram of an ion mobility spectrometer. The typical IMS consists of an open tube in which an electrical field has been place in one direction and a flow of "drift" gas is flows counter to the electric field. In the diagram in Figure 1a, an ionization source is depicted on the left of the Figure. Traditional IMS systems this ionization source is a radioactive ^{63}Ni source which produces both negative and positive ions from the air or nitrogen drift gas. These ions, produced from the drift gas, are reacted with dopant gases such as ammonia, acetone, or methylene chloride to produce reactant ions. In the presence of a chemical warfare agent, these reactant ions undergo ion-molecule reactions to produce a product ion which is indicative of the chemical warfare agent. Other ionization methods such as photoionization, corona ionization, electrospray ionization and secondary electrospray ionization are also used to produce product ions of the CWAs.

Once the agent enters the detector and corresponding product ions have been formed, they migrate through the gas at atmospheric pressure in the direction of the electric field. If the field is positive then the positive product ions will migrate from left to right the spectrometer depicted in Figure 1. If the field is negative, then negative ions will migrate toward the detector. Electropositive agents will tend to form positive ions with the addition of a proton while electronegative agents will tend to capture an electron to produce negative product ions. The product ions migrate in the field toward the detector until the reach the ion gate where they are neutralized and swept out of the instrument with the opposing drift gas flow.

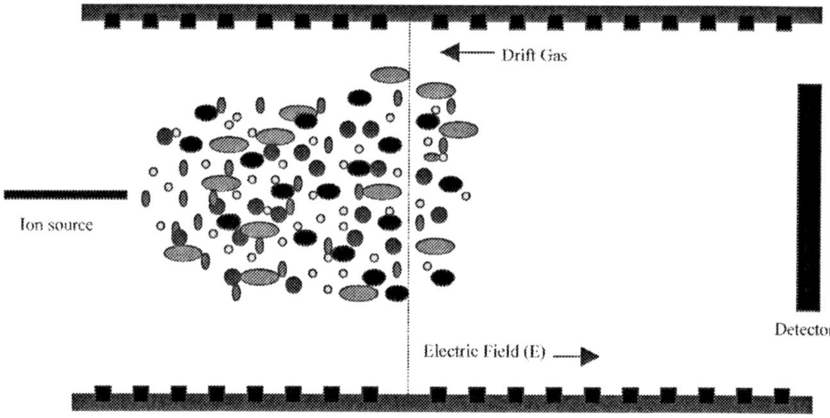

Figure 2: Ion injection

Periodically, as shown in Figure 2, the ion gate is opened and a pulse of ions is permitted to pass through the gate into the ion separation region. This ion injection process is typically repeated at a frequency of 20 Hz. The pulse width of the ions injected can range from 50 to 500 μsec but is typically around 200 μsec.

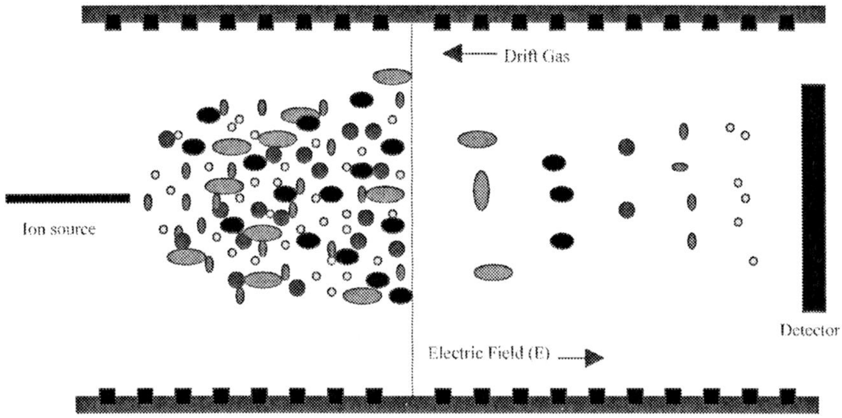

Figure 3: Ion Separation

After the ion pulse has been injected, the gate closes and the ions continue to migrate toward the detector and separate according to their differing mobilities as shown in Figure 3.

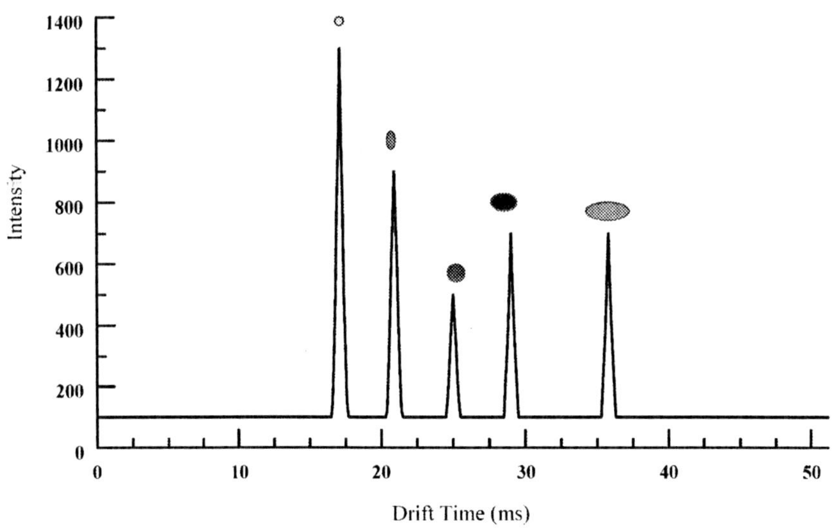

Figure 4: Ion mobility spectrum

Smaller ions have smaller collision cross sections with the drift gas and therefore migrate faster in the gas. Thus, ion mobility spectrometry is a size separation process. Additionally, smaller ions migrate faster than larger ions containing the same charge. As the ions impinge on the detector, they are neutralized, causing current to flow which is recorded at the arrival time as an ion peak to produce an ion mobility spectrum as shown in Figure 4.

Arrival times or drift times are typically on the order of milliseconds. By knowing the length of the drift region and the time that the ions required to traverse the drift region, the ion's velocity can be determined. Under conditions normally used for IMS when detecting CWAs, the ion velocity is directly proportional to the electric field.

$$\upsilon = KE \Rightarrow \frac{L}{t_d} = K\frac{V}{L} \Rightarrow K = \frac{L^2}{Vt_d}$$

Where v is the ion's velocity, E is the electric field, V is the voltage across the drift region, L is the length of the separation region, t_d is the arrival time of the ion and K is the mobility constant. The mobility constants must be corrected for temperature and pressure as shown below:

$$K_o = K\frac{273}{T}\frac{P}{760}$$

Where K_o is the reduced mobility constant and P and T are the operating pressure and temperature, respectively. When the mobility constant is corrected for temperature and pressure, the reduced mobility constant becomes a qualitative identifier of the ion and thus, the chemical agent.

Ion mobility spectrometers can be small and light. One IMS instrument was described during the presentations at the NATO-Russia Workshop on Ecological Risks Associated with the Destruction of Chemical Weapons. It was called a μIMS and was produced by G.A.S. in Dortmund, Germany. The resolving power of this hand held IMS instrument was reported to be 30-60 using a ^{63}Ni source for detecting volatile organic compounds in the low ppm to low ppb range. It was programmable with an RS 232 interface and could be battery operated for up to 6h.

Recently, atmospheric ion mobility spectrometers have been shown to exhibit high resolution separations. Table 1 compares IMS resolving powers with that of other chromatographic methods of separation. Traditionally IMS resolving power has been measured by taking the ratio of the arrival time to the ion pulse width at half height (t_d/W_h). The highest resolving power which has been recorded by IMS is 150, which corresponds to 130,000 theoretical plates in chromatographic terminology. As can be seen from the table, IMS has a resolving power greater that HPLC and similar to capillary gas chromatography. More importantly is that its time of separation in on the order of 50 ms, which means that it produces over 2 millions plates per second compared with 14 for HPLC, 100 for capillary GC and 500 for capillary electrophoreses. This makes IMS an ideal separation method to interface to mass spectrometry for two-dimensional analysis.

Table 1: IMS vs. Chromatography

IMS vs. Chromatography

	HPLC	GC	CE	High Res. IMS
# of Theoretical Plates (N)	25000	120,000	300,000	130,000
Efficiency (H.E.T.P)	10μm	400μm	2μm	2μm
Resolving Power (R_p)	65	145	230	150
Time of Separation	30 min.	20 min.	10 min.	50msec
Plates per Second	14	100	500	2,600,000

4. TWO-DIMENSIONAL SENSORS

Two dimensional sensors add a second, orthogonal mode of separation to further increase selectivity and decrease the potential of a false positive response. Common two dimensional sensors include gas chromatography/mass spectrometry (GC/MS), liquid chromatography/mass spectrometry (LC/MS), capillary electrophoresis/mass spectrometry (CE/MS) and ion mobility/mass spectrometry (IMMS). Table 2 lists several of the traditional 2-D methods for monitoring CWAs along with their advantages and disadvantages.

The GC/MS sensor is the only sensor which is approved for use with the Chemical Weapons Convention although others are used routinely for other applications of monitoring CWAs and their degradation products. GC/MS is of course limited to monitoring volatile chemicals while LC/MS, CE/MS are used for non-volatile compounds. One primary advantage of IMMS is that it can be used to monitor both volatile and non-volatile chemicals.

Table 2: Multidimensional Sensors

Method	Advantages	Disadvantages
HPLC with fluorescence or UV detection	1. Adaptable to field analysis 2. Laser induced fluorescence provides excellent sensitivity	1. Derivatization required 2. Long analysis times 3. Insufficient sensitivity with most detection methods
HPLC with MS detection	1. Sensitivity 2. Positive identification	1. Not field applicable 2. Long analysis times
GC with MS detection	1. Positive identification 2. Fast analysis times 3. Sensitivity	1. Derivatization required
µ-LC with Flame Photometric Detection (FPD)	1. Fairly Sensitive 2. FPD responds to most nerve and blister analytes	1. Requires sample pretreatment, cleanup, and enrichment 2. Not field developed
CE and CE with UV or ESI-MS detection	1. MS provides positive identification 2. Excellent separation 3. MS sensitivity	1. Long separation times 2. Often requires sample pretreatment 3. Has only been applied to a narrow range of analytes

Figure 5 shows a picture of a prototype ion mobility (time of flight) mass spectrometer used for the 2-D detection of chemical warfare agents and their degradation products. Degradation products are particularly difficult to monitor by GC/MS because many of these compounds are polar, water soluble and non-volatile. The electrospray ion source is shown on the far right of the picture. Aqueous samples of non-volatile chemicals can be introduced through the electrospray needle directly into the IMS, forming product ions directly from the spray process. For volatile compounds the reactant ions produced by the electrospray source can be used to produce product ions from the gas phase neutrals through ion-molecule reactions similar to those which occur in radioactive and corona ionization sources. The ion mobility spectrometer is shown in the middle of the picture and the time-of-flight mass spectrometer is seen orthogonally positioned to the IMS and at the left of the picture.

Figure 5: Electrospray Ionization – Ion Mobility – time-of-flight – Mass Spectrometer

The advantages of IMMS for the detection of chemical warfare agents and related compounds are many: The IMS serves as an interface to the mass spectrometer, keeping it clean and reducing the MS noise which can reduce sensitivity and lead to false negative responses. With 2-D high resolution separation, identification of chemicals is more reliable with fewer false positive responses. The IMS interface to the MS allows the instrument to be used to monitor both air and water samples, volatile and non-volatile chemicals. While the current prototype is not field-portable, the current model could be incorporated into a 2'x 2' x 2' box for transportation and field operation. Future designs could be much smaller.

Overall, the goal of having a light-weight, low-power, inexpensive and highly dependable monitor for CWAs is not currently possible. But the resolving power to instrument size ratio is increasing every year and with each new instrumental development. Thus, the goal of achieving a "tricorder" like sensor no longer seems impossible in the not too distant future. Until that time however, current instruments, complete with all there disadvantages must be employed in the field for real time screen devices, followed by confirmatory laboratory analyses.

References

[1] Chemical Weapons Convention (CWC) bans the development, production, acquisition, stockpiling, and use of Chemical Weapons and on there destruction. Washington D.C. United States Bureau of Arms Control and Disarmament Agency, Entered into force April 29, 1997

[2] Hill, H.H.; Martin, S., *Pure and Appl. Chem.* 2002, 74, 2281-2291

IDENTIFICATION OF CHEMICAL NEUTRALIZATION OF PRODUCTS OF ORGANOPHOSPHORUS CHEMICAL WARFARE AGENTS

Elena Saveljeva[1], Igor Zenkevich[2], Andrey Radilov[1]
[1]Research Institute of Hygiene, Occupational Pathology and Human Ecology St.Petersburg (Russia), [2]Chemical Research Institute of St.-Petersburg State University, St.Petersburg (Russia)

Abstract: Investigation of the bitumen salt masses (BSM) ingredients is required to evaluate toxicity and hazard of chemical weapons destruction products. The identification of components is feasible in the case of simultaneous consideration of mass spectrometric and chromatographic findings with the use of all known methods of calculation of the retention indices. The combined interpretation of parameters obtained within the frames of only one GCMS-EI instrumental method can be no less efficient than the combination of two and more different instrumental methods.

Keywords: bitumen salt masses, organophosphates, mass spectrometry, chromatography

In the Russian Federation, chemical neutralization with alkaline compositions containing alcohols, N-methylpyrrolidone or monoethanolamine is used to destroy organophosporus chemical warfare agents (CWA). One of the alternative methods of utilizing products of CWA chemical neutralization involves their bituminization to produce bitumen salt masses (BSMs). Investigation of the BSM ingredients is required to evaluate the toxicity and dangerousness of products issued from the destruction of chemical weapons. In view of the fact that non-volatile components are dominant among decomposition products of organophosphorus CWA, the trend toward their analysis by methods of liquid chromatography and capillary electrophoresis is evident. Electron impact (EI) mass spectra of the vast majority of these products have low information content, signals of molecular ions being frequently absent in these spectra. This being so, mass spectrometric identification of products of organophosphorus CWA decomposition are principally performed in chemical ionization and tandem modes. To detect the structure of unknown chemical compounds, the use of FTIR spectroscopy, atomic emission analysis, and other methods is justified along with mass spectrometry. The combination of the above methods of separation and detection is feasible with the use of up-to-date expensive instrumentation equipment. The urgency of these problems justifies the high

V.M. Kolodkin and W. Ruck (eds.),
Ecological Risks Associated with the Destruction of Chemical Weapons, 167–172.

cost of methods of solution, whereas potentials of moderately priced instruments providing only gas chromatographic (GC) separation and mass-selective EI detection (GCMS-EI) are far from being exhausted.

This paper presents findings obtained as a result of GCMS-EI identification of organophosphorus CWA chemical neutralization products in BSM compositions. All the investigations were carried out using QP 5000 instrumentation unit of the Shimadzu Company (Japan) with the use of the capillary column with DB-5 stationary phase in the temperature programming mode, with recording standard mass-spectra and measuring linear-logarithmic retention indices (RI) in the range from 600 to 2600.

The identification of components is feasible only in the case of simultaneous consideration of mass spectrometric and chromatographic findings with the use of all known methods of calculation of the retention indices and interpretation of these parameters from the structural point of view.

Volatile organic compounds were analyzed in gas evolutions from BSMs at the melting temperature of samples by means of head space technique **(Method I)**. Under these conditions alcohols, dialkyl esters and thioesters of methylphosphonic acid, products of organophosphorus CWA degradation including those of the VX group belonging to diethylaminoethylsulfides and polysulfides, as well as diethylamine, diethylformamide, N-methylpyrrolidone (neutralizing agent), caprolactam (viscosity adjuster), and other compounds were detected. Non-volatile organic compounds were detected in dichloromethane extracts from BSMs after their re-solution in acetonitrile and silylation with bis(trimethysilyl)trifluoroacetamide (BSTFA) **(Method II)**. Methylphosphonic acid and its acidic esters, as well as mono- and diethanolamine, were analyzed as trimethysilyl derivatives.

The exception is S-(2-diethylaminoethyl)methylphosphonic acid $MePO(OH)SCH_2CH_2NEt_2$ that, according to prediction, is the most toxic compound of all the products identified. This compound is not only not detected by direct gas chromatographic analysis, neither can it be silylated because of the zwitterion formation. Owing to this, S-(2-diethylaminoethyl)methylphosphonic acid was analyzed as a methyl derivative after treating with trimethylanilynium hydroxide with the resulting formation of ion pairs that were broken down at high temperature in the injector where the acid is methylated **(Method III)**.

S-(2-diethylaminoethyl)methylphosphonic acid, the decomposition product of "Russian VX", is a structural analog of the highly toxic substance EA 2192 formed when common or "American VX" decomposes.

Russian VX,

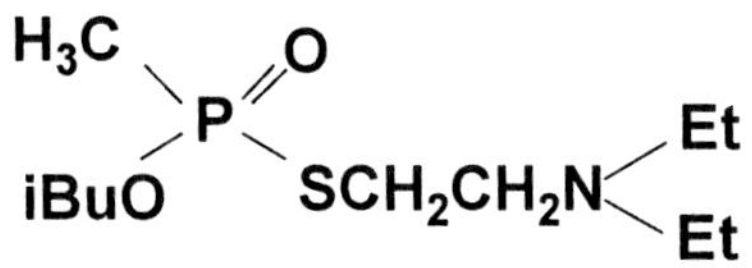

was produced only in the former Soviet Union. There were no data on products of its transformation available in literature. For American VX,

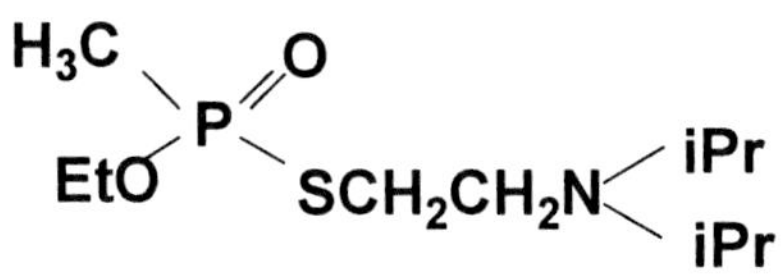

not only some products of its decomposition were known but also their RI values have been published [1]. We used these values as base units to predict RI values for unknown RVX transformation products, using additive schemes. RI experimental values have been determined and EI mass-spectra have been obtained for 43 components of the reaction mixture formed after neutralizing RVX. By correlating RI experimental and estimated values with the simultaneous interpretation of EI mass-spectra 24 components of 43 have been identified. The estimation algorithm and a logical scheme of identification are comprehensively discussed in [2 - 4].

Since the vast majority of investigations involving the identification of products of organophosphorus CWA transformations have been conducted and, evidently, will be conducted in future by hyphenated techniques with the use of gas chromatography for separation, RI values remain the most universal and undeniably necessary means to solve these tasks. On the basis of our experience, we suggest that the combined interpretation of parameters obtained within the frames of only one GCMS-EI instrumental method can be no less efficient than the combination of two or more different instrumental methods.

As to a series of products of organophosphorus CWA transformations, particular attention, in our opinion, should be paid to substances retaining C-P bond in a molecule. These compounds can be selectively detected in various matrices and used as indicators or markers that allow the conclusion to be made concerning pollution of environmental subjects and areas with decomposition products of organophosphorus CWA.

The Table presented lists markers of organophosphorus CWA identified in BSM samples containing GB-, GD-, and RVX destruction products.

Table 1: Chemical Markers for Retrospective Analysis of Organophosphorus CWA

N	Compound	RI (DB-5)	$(m/z)_{max}$	Method of analysis
1	2	3	4	5
1	$PO(CH_3)(O\ iPr)_2$	1067	97	I
S1**	$PO(CH_3)(OH)(O\ iPr)$	1110	153	II
2	$PO(CH_3)(O\ iPr)(O\ iBu)$	1142	97	I
S2**	$PO(CH_3)(OH)_2$	1147	225	II
S3**	$PO(CH_3)(OH)(O\ iBu)$	1221	153	II
3,4*	$P(CH_3)O(O\ iPr)(O\ Pin)$	1270 1275	97	II
5	$PO(CH_3)(O\ iBu)_2$	1298	97	I
6	$PO(O)(O\ iPr)_2(O\ iBu)$	1317	99	II
S4**	$PO(CH_3)(OH)(O\ Pin)$	1320	153	II
7	$P(CH_3)S(O\ iBu)_2$	1331	113	II
8	$P(CH_3)S(O\ iPr)(O\ Pin)$	1347	113	II
9, 10*	$P\ O\ (CH_3)\ (O\ iBu)(O\ Pin)$	1393 1402	97	II
10	$PO(CH_3)\ (OH)SCH_2\ CH_2N\ Et_2$ (O-methyl derivative)	1424	86	III
11	$PO(CH_3)(O\ Pin)_2$	1501	97	II
12	$PO(CH_3)\ (S\ iBu)SCH_2\ CH_2N\ Et_2)$	1782	86	II
13	$PO(CH_3)\ (O\ iBu)(SCH_2\ CH_2)_2N\ Et_2$	2128	86	II
14	$PO(CH_3)\ (O\ iBu)(SCH_2\ CH_2)_3N\ Et_2$	2636	86	II

* two diastereomers
** TMS esters

Methylphosphonic acid and its acidic esters analyzed as trimethylsilyl derivatives and listed in the Table as S1 – S4 were isolated from blood and urine of individuals exposed to terror attacks using CWA in Japan (Matsumoto, Osaka, Tokyo 1994-1995). At present it has been found that these compounds are biological markers, detection of which in human biological media is a reliable indicator of the body's exposure to CWA. Devising methods for the analysis of CWA biological markers in human blood and urine is needed to observe the state of health of persons working at chemical weapons destruction facilities, as well as in diagnostic and forensic purposes in cases of anticipated acts of terrorism using CWA.

The analysis of methylphosphonic acid and its acidic esters in biological media requires solving two problems, namely - destruction of adducts of these compounds with proteins and overcoming the matrix effect that retards silylation and chromatographic separation of the sample.

To solve the first problem, the use of mild hydrolysis with the participation of alkaline phosphatases is preferable. Today, there is no adequate solution to the problem of the matrix effect elimination. Various concentrating and cleaning up procedures with the use of solid-phase extraction methods are at the point of approbation. The method based on the combination of sorption and ion-exchange procedures seems the most efficient.

References

[1] D'Agostino P.A., Provost L.R., Visentini J. J. Chromatogr. 402 (1987) 221.

[2] Saveljeva E.I., I.G.Zenkevich and A.S.Radilov. Identification of O-Isobutyl S-(2-Diethylaminoethyl) Methylthiophosphonate Chemical Neutralization Products in Bitumen-Salt Matrices. J. Anal. Chem. 2003, V.58, N 2, P. 135-145.

[3] Saveljeva E.I., Zenkevich I.G., Radilov A.S. Identification of V-type Chemical Warfare Agent (Russian VX) Destruction Products with GC-MS. 25th International Symposium on Capillary Chromatography. Riva del Garda, Italy. May 13-17, 2002. N 09. CD ROM.

[4] Saveljeva E.I., Zenkevich I.G., Radilov A.S. GCMS helps to chemical disarmament. Shimadzu NEWS. 2003 N 1, p.6 –8.

ENVIRONMENTAL AND HEALTH MONITORING IN RELATION TO THE DEMOLITION OF THE FORMER CWPF AT JSC KHIMPROM, NOVOCHEBOKSARSK, RUSSIA

Tatyana G. Kudankina[1], Nina P. Pavlikova[2], Martin Silberschmidt[3], Lars Carlsen[3*]

[1]*Center for State Sanitary Supervision № 29, Novocheboksarsk (Russia),* [2]*Analytical Laboratory of Shop 83, Detached Plant № 4 of JSC Khimprom, Novocheboksarsk (Russia),* [3]*Tacis Chemical Weapons III Project, Novocheboksarsk (Russia).*

Abstract: The environmental and health monitoring activities associated with the planned demolition of the former chemical weapons production facility (CWPF), the detached Plant No 4 at JSC Khimprom, Novocheboksarsk, Russia are presented, the prospected demolition being briefly outlined. The various organizations responsible for environmental and health activities are introduced. Through a comprehensive environmental impact assessment, the major risk factors for the workers, the population and the environment has been outlined. It is concluded by the environmental impact assessment that, if the accepted solutions are followed and observed, the demolition of the former CWPF will not significantly affect the environment and the living conditions of the population of Novocheboksarsk and the surrounding territories. A close collaboration between national and regional socio-hygienic monitoring activities and local information activities are strongly emphasized in order to achieve maximum public confidence and acceptance in the demilitarization process.

Key words: CWPFs, Demolition, Health Monitoring, Environmental Monitoring, Integrated Approach

1. INTRODUCTION

In the past decades, the world community has been trying to eliminate chemical weapons (CWs) as weapons of mass destruction. One of the most significant results of these efforts was the development of the 'Convention for the Prohibition of the Development, Production, Stockpiling and Use of

* Corresponding author; permanent address: Awareness Center, Hyldeholm 4, Veddelev, DK-4000 Roskilde, Denmark; e-mail: LC@AwarenessCenter.dk

V.M. Kolodkin and W. Ruck (eds.),
Ecological Risks Associated with the Destruction of Chemical Weapons, 173–197.

Chemical Weapons and their Destruction' (CWC) [1] that entered into force on April 29th, 1997. CWC gives major emphasis on the declaration and destruction of existing stockpiles of CWs. The Russian Federation ratified the CWC in November 1997. Subsequently, Russia, as a major possessor of CWs, stated it had started with the destruction of its stockpiled chemical weapons. Thus, by April 2003, Russia successfully met the 1% deadline for destruction of category one chemical weapons [2].

Further to the destruction of existing chemical weapons the CWC covers the destruction or conversion of chemical weapons production facilities (CWPFs) [1]. With reference to the CWC Russia declared, among other plants, that the detached Plant № 4 at JSC Khimprom, Novocheboksarsk, as a former CWPF for VX[1] in 1998, and its demilitarization was initiated as early as 1990 [3, 4]. Thus, according to Article IV of the CWC, this facility should be demolished 10 years after entry into force of the CWC for Russia at the latest. VX production at Plant № 4 was in operation from 1972 until 1987. In 1987 the production had been stopped and the building had been mothballed. A plan containing a list of the parts to be demolished and the methods to be applied has been submitted to the Organisation for the Prohibition of Chemical Weapons (OPCW) [4] in 2000 by the Russian Federation. Between 1997 and the submission of the detailed plan for destruction of the former CWPF in 2000, dismantling activities continued and have been reported to the OPCW. The former CWPF went through initial inspection by OPCW inspectors in July 1998 [5]. The Executive Council of the OPCW has subsequently approved the plans for destruction of Plant № 4 in May 2000 [4a].

In accordance with the CWC, the member state must give priority to the safety of people and to the protection of the environment by destruction of CWs (Article IV paragraph 10) and former CWPFs (Article V paragraph 11) according to national standards and regulations. The basis for developing a national strategy concerning the safe destruction of chemical weapons as well as former CWPFs is thus emphasized by the Russian laws related to public health as well as to environmental protection [3, 6a, 7-9]. The safety of both the personnel directly engaged in the proposed operations at the facility and the population of the adjacent communities as well as the environment is obviously an issue of major importance.

The Tacis project "The development of an Environmental and Health Monitoring System related to the Demilitarisation of the Chemical Weapons Facility at detached Plant № 4 of JSC Khimprom, Novocheboksarsk" [10]

[1] The Russian version of VX: *S*-2-(diethylamino)ethyl *O*-(2-methylpropyl) methylphosphonothiolate; CAS Registration № 159939-87-4

was initiated in January 2003. The project will be finished by August 2004. The project focuses on supporting the existing monitoring systems. Thus, the focus is on analyzing the existing monitoring systems, their applicability for the planned demolition work and possibly suggesting effective ways to further improve the system sustainability. The basic component of the project is the strengthening of the technical capabilities of the institutions to be involved in the monitoring activities during the demolition phase through installation of specific analytical equipment to sustain the health and environmental monitoring activities, respectively, as well as industrial and socio-hygienic monitoring. A further part is the establishment of an information and communication system.

The present paper describes the joint effort of the involved Russian laboratories and the Tacis project to establish an integrated health and environmental monitoring system that eventually will be further integrated with appropriate information activities thus leading to an integrated Health and Environmental Monitoring and Information System (HEMIS).

2. DEMOLITION WORK TO BE PERFORMED

Under the plan for demolition of the former CWPF "Detached Plan № 4 of JCS Khimprom" it is foreseen that the building, which housed the main production of VX and filling of munitions will be destroyed latest until 2007 [4]. Thus, the detailed plan approved by the OPCW stipulates consecutive steps on the demolition, the methods used as well as the schedule of the destruction of different equipment and the building itself [4]. Parts of the equipment in the plant have been demilitarized prior to entry into force of the CWC (*vide supra*). In the following a brief outline of the projected demolition process is given.

2.1 Brief outline of the projected demolition process

The demolition of the former CWPF is planned to consist of three stages. The first phase comprises the demilitarization of the so-called clean facilities originally used for the production of intermediates for VX production – buildings 350A, 370, 370b, an assembly line to equip the filled munitions with explosive shells – buildings 1, 3, and the first stage of filling the munitions, i.e. preparation for filling in building 351 with its galleries (cf. Fig. 1). This phase of the demilitarization of the facility was completed by December 2002.

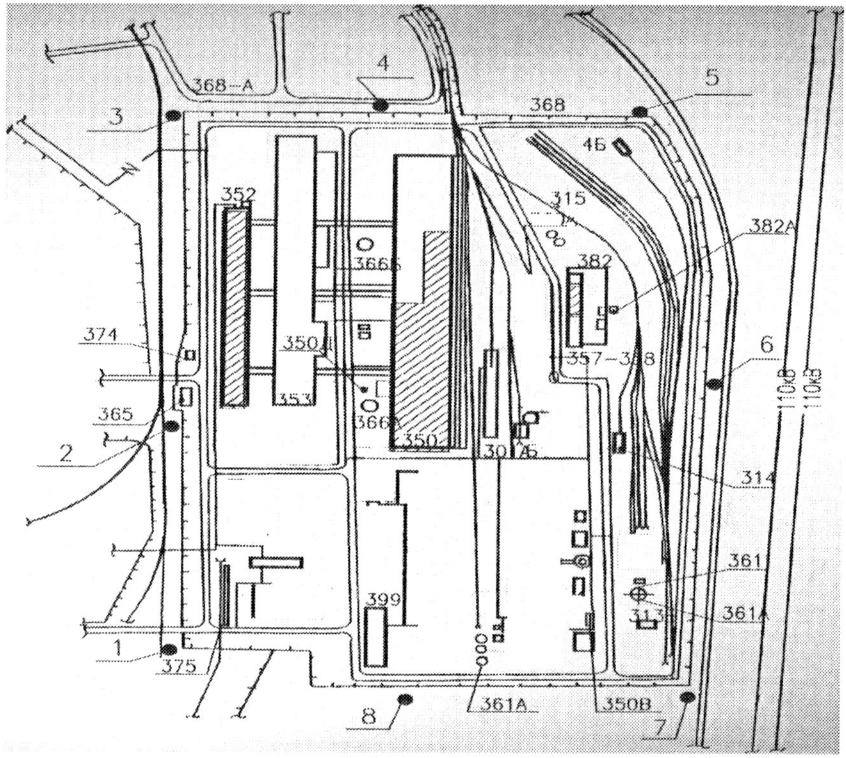

Figure 1: Location of the single buildings at the Detached plant № 4 at JSC Khimprom, Novocheboksarsk.

The second phase of the demilitarization will include two steps. The first step includes a) the development of a base for the physical demolition and removal of signs of the former VX production in buildings 350, 352, 353, 366A (cf. fig 1), b) the installation of an incinerator devoted to the thermal treatment of liquid wastes from the demilitarization of buildings 350, 382, 382A (cf. Fig. 1) and c) the installation of two thermal treatment units to burn solid wastes. The second step is the decontamination and decommissioning of specialized equipment in buildings 350, 352, 353 (cf. Fig. 1).

The second phase of the demilitarization will be conducted consecutively and will include three stages in addition to the preparatory phase of stage I.

The preparatory phase of stage I comprises repair and reconstruction operations of building 350. This includes ventilation systems, sewage, energy supply networks and technological circuits to provide for the demilitarization, as well as the second check of the waste burial ground (disposal burial check 2). In relation to building 352, the preparatory phase comprises a set of operations related to the repairs of the building itself, reconstructing the lighting, heating, communication, alarm, ventilation systems, reconstruction of the sanitary check point, personal protective equipment decontamination unit, laboratories.

Stage I comprises the setting up of a base for waste processing, decontamination of the equipment of stage II and III dismantling and cutting of the equipment, metal parts, air ducts, setting up partitions, building solid waste and waste-water thermal units, regular operation of all technological sections of building 352.

Stage II includes all kinds of decontamination operations, cutting and thermal treatment of solid and liquid wastes as well as regular operation of all the technological sections of building 352.

During stage III, the final demilitarization operations in building 352 are completed while decommissioning building 350. The laboratories of building 352 will be engaged in monitoring the wastes from the demolition of buildings 350, 366A and the galleries.

The third and final phase of the demilitarization comprises the eventual dismantling and disposal of the construction structures of building 350, galleries 3, 4, 5, and stack 366A (cf. Fig 1).

3. PLANNED MONITORING ACTIVITIES DURING THE DEMOLITION PHASE

During the production period, 1972-1987, VX was identified as the most hazardous chemical compounds being present at the facility. Due to the very low volatility of VX (vapor pressure at 25°C: $2.5\cdot10^{-4}$ mmHg [11]) and the fact that VX hydrolyses only slowly under humid conditions, VX appears to be rather persistent in the environment. Thus, at pH < 6 the half life of VX[2] in sea water or river water appears to be 1-2 years. In the terrestrial environment, the persistence ranges from 10 days in grass/woodland at +15°C to 8 weeks in areas covered with loose snow at -10°C [12]. Thus, the VX production prompted, due to the extreme biological activity of the compounds used and produced, the development of a protection system to safeguard the population of the adjacent areas. The system is aimed at

[2] Data on hydrolysis and persistence corresponding to the American VX (*S*-2-(diisopropylamino)ethyl *O*-ethyl methylpropyl) methylphosphonothiolate; CAS Registration № 50782-69-9) [8]

preventing a possible impact of hazardous factors on public health and the environment.

It should be emphasized that throughout the whole period of the production monitoring (1972-1987), VX was detected neither in the ventilation exhausts, on the territory of the industrial site, on the border of the sanitary protection zone, in the city of Novocheboksarsk, nor in the soil and water reservoirs [11]. Later studies in 1993 and 1998 confirmed the absence of VX contamination of the waters and bottom sediments of the open water reservoirs and soils within 30 km from JSC Khimprom as well as in atmospheric air on the border of the sanitary protection zone [13].

The VX production at the Novocheboksarsk facility was only a fraction of the rather diversified chemical enterprise Khimprom. Thus, comprehensive monitoring of atmospheric air and other environmental media has been, and continuously is, being conducted for all major types of emissions from the various productions. The major air pollutants have been identified as ammonia, hydrogen chloride, chlorine, chlorophos, aniline, chlorinated hydrocarbons etc [13]. VX contamination has never been detected.

It is noteworthy that, in addition to the former production of VX and the various productions at JSC Khimprom in general, emissions from the nearby Power Plant 3 must also be taken into account when evaluating the impact on the environment and thus the health of the population living within 4.5-7 km of the industrial area including both JSC Khimprom and Power Plant 3. In this connection, it should be noted that the pollution detected apparently does have a certain impact on public health. Especially the impact on the health of the children, being the most vulnerable part of the population has been observed. Consequently, these investigations led to activities and subsequently to actions to decrease the pollution from ventilation exhausts. Furthermore, the atmospheric air pollution monitoring was strengthened.

Monitoring of VX has, up until now, been performed by a biochemical method (cholin esterase activity tests). In comparison to modern, e.g., chromatography-based analytical techniques utilizing specific detector systems, these assays appear to be rather unspecific and inaccurate. Thus, part of the Tacis project will be to introduce proper analytical equipment in the laboratories responsible for the monitoring activities.

The impact of the demolition process may be caused by the actual presence of the residual amounts of VX in the industrial building 350 (cf. Fig. 1). Further, the possible emissions of general industrial contaminants to the environment through atmospheric emissions and through liquid wastes, as well as the possible effects of physical factors (noise, heat, electromagnetic radiation etc.) appear as major potential occupational environment impact factors.

Estimates on the potential impact on the environment and on human health of the demilitarization operations are comprehensively presented in the Environmental Impact Assessment [11]. It has been stated [11] that the demolition workers health risk is mostly related to skin contamination. Thus, this group will be submitted to a specific health-monitoring program during and after the demolition period.

Where the surrounding environment is concerned, the expected predominant impact is caused by contaminants emitted into the atmosphere. The other environmental compartments will be affected only indirectly through atmospheric depositions [11]. Up to 24 substances are expected to be emitted into the atmosphere during the demolition. However, only 2 of those, i.e., chromium oxide and ozone refer to extremely hazardous compounds, whereas the other ones refer to highly hazardous (7 compounds), moderately hazardous (6 compounds) and low hazardous (3 compounds) type. Six compounds including VX have not been assigned a hazard rank. Thus, their acceptable levels in the environment are being determined by a Predicted No Effect Level (PNeL) as a hygienic standard.

The Environmental Impact Assessment [11] concludes that if the accepted solutions are followed and observed the demolition of the former CWPF, detached Plant № 4, at Khimprom, Novocheboksarsk, will not significantly affect the environment and the living conditions of the population of Novocheboksarsk and the surrounding territories within a distance of 2-4 km from the facility.

3.1 Environmental monitoring

Within the region that may possibly be contaminated as a consequence of the planned demolition of the former CWPF, Plant № 4, a comprehensive system of industrial environmental monitoring (IEM) has been established [14]. The IEM system operates as a multilevel system providing valid and reliable information on the contamination of all the areas of the environment affected by emissions from JSC Khimprom.

The IEM system envisions that the JSC Khimprom industrial laboratory continuously monitors general industrial pollutants originating from the ordinary production. Thus, the following media are continuously monitored (cf. Fig. 2):

- atmospheric air on the industrial site, in the sanitary zone and in surrounding residential areas
- atmospheric emissions from specific emission sources at JSC Khimprom
- wastewater sent to the neutralization station and subsequently processed to the biological treatment plant

- surface water at the point of discharge, in the reservoir water and at the sampling points both up and downstream from the discharge point
- groundwater from observation wells on the industrial site as well as on the waste disposal sites
- soil at the border of the JSC Khimprom sanitary zone and at the waste burial ground

The monitoring system is organized and operated in accordance with legislation related to this type of activities that is in effect on the territory of the Russian Federation. The state supervision is conducted by the Center for State Sanitary Supervision № 29 of the Medbioextrem Federal Agency, Russian Health Ministry, the Specialized Inspection of Analytical Control, SIAK, of the Volga Region, the Chuvash Republic Branch of the Ministry of Natural Resources and the Chuvash Republic Hydrometeorology and Environmental Monitoring Center.

The Environmental Impact Assessment [11] states that the major impact on the environment during the demolition will be due to atmospheric emissions of pollutants. The water and soil environmental compartments were assessed to be affected only indirectly by atmospheric deposition [11].

Wastewater from the decontamination activities that might contain VX will be directed to thermal treatment in the designated incinerator (vide supra). Thus, discharge to the biological treatment plant is excluded. VX contaminated solid wastes shall undergo thermal treatment as well. Thus, disposal of untreated wastes is excluded.

During the demolition operations the existing monitoring system is not supposed to undergo any significant changes except for the fact that all environmental elements, including snow, as well the emission sources of the former CWPF, will be subject to specific monitoring for VX. It further appears that the present network of sampling points for all environmental is virtually sufficient to fulfill the needs during the demolition process (cf. Fig. 2 and 3). However, two additional observation wells will be installed on the territory of the former CWPF. Apart from these two additional wells, no further sampling and observation points will be installed. Thus, during the demolition process of the former CWPF the existing monitoring system will allow for tracing VX on site, on the border of the JSC Khimprom sanitary zone, and on the adjacent territory.

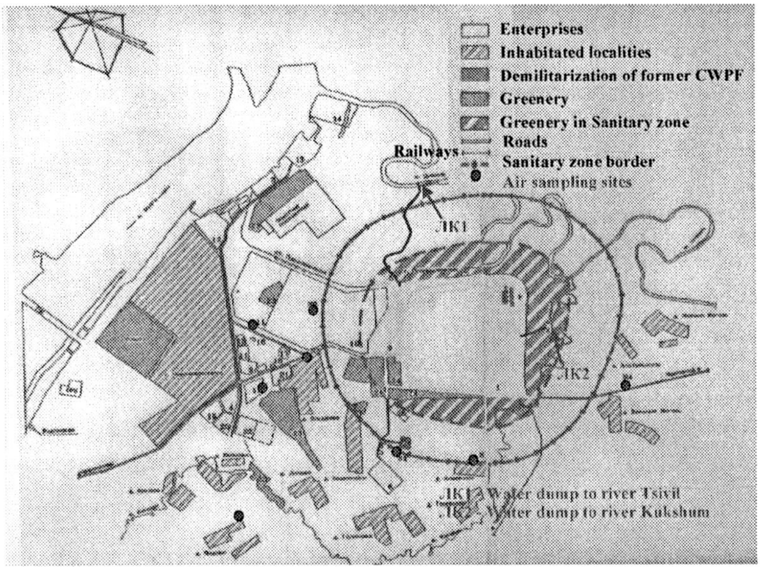

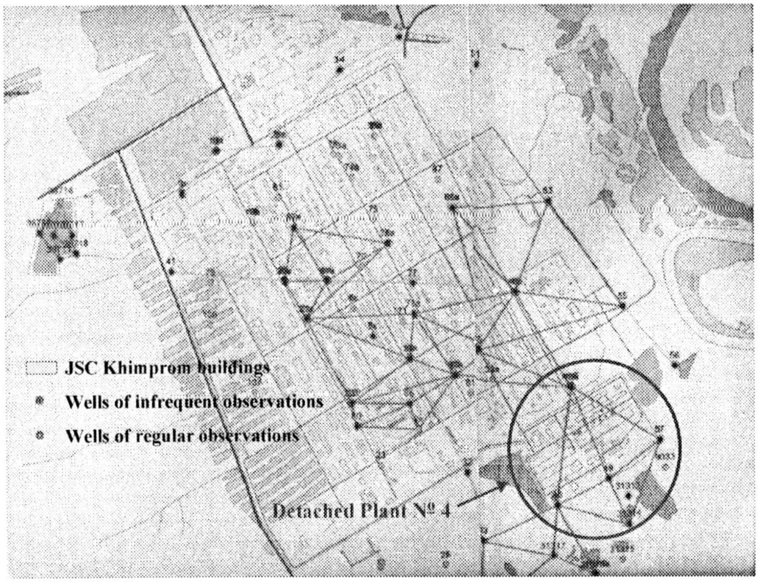

Figure 2, 3. Location of ground water sampling points at JSC Khimprom, Novocheboksarsk

In relation to the possible improvement of the existing monitoring program through the Tacis project, the starting point has been the monitoring plan developed for the demolition process [14]. Further, baseline environmental monitoring studies are based on a series of air, soil and water samples that is currently being sampled inside, as well as outside, the sanitary zone of JSC Khimprom. The samples comprise 8 air, 8 soil samples, 3 surface samples and 4 groundwater samples all in duplicate. Thus, in total 46 samples will be analyzed. The analyses of these samples will be carried out at GosNIIOKhT by laboratory staff from the State Center for Sanitary Supervision № 29 and Analytical Laboratory of Workshop 83, Detached Plant № 4 of JSC Khimprom, both Novocheboksarsk, Russia, as an integrated part of the training activities for laboratory staff as foreseen by the project.

3.1.1 Improving analytical capabilities

An important part of the Tacis project is the supply of the analytical laboratories with suitable equipment for analyzing the hazardous material possibly being emitted during the demolition, primarily VX. Thus, providing equipment is mandatory in order to ensure high quality quantitative chemical analyses of VX. As mentioned above, the previously applied methods based on the choline esterase assays are substituted by chromatographic methods using various specific detectors such as flame photometric and thermo ionic detectors, specific for phosphorous and sulfur detection and mass selective detections. This equipment will ensure high sensitivity of the VX analyses with maximum validity and selectivity.

The Analytical Laboratory at Shop 83 and the Center for State Sanitary Supervision № 29 will conduct the VX monitoring, at the detached Plant № 4. The sampling of environmental samples for environmental monitoring as well as the analytical work will be carried out according to Russian GOST Standards and thus complies with Russian legislation.

The methods to be used for the analyses of VX in various media have been developed by GosNIIOKhT, who has a considerable experience in dealing with chemical agents and is accredited by the state standardization body Gosstandart to perform this kind of activities. Available methods have a necessary sensitivity of 0.5 MAC of VX or less and have passed the metrological attestation with the Gosstandart.

For the time being, the Detached Plant № 4 of JSC Khimprom has available the following chromatographically based methods for determining VX in various media:

a) In accordance with hygienic regulations:

- occupational air
- surfaces of the technological equipment
- surface waters
- soil

b) In accordance with technological regulations:
- construction materials
- used decontamination formulations
- cleaning water

Presently, the Detached Plant № 4 does not have the following methods of determining VX available:
- furnace sludge (slurry)
- ventilation emissions
- air
- surfaces of the personal protective equipment
- surface of the construction materials

GosNIIOKhT is currently developing and documenting methods for the analysis of VX in these media.

The implementation of these analytical techniques, as well as the necessary training of the laboratory staff are integrated parts of the improvements of the existing environmental monitoring system. It is foreseen that the installation of the equipment and the corresponding trainings activities will be completed well in time before the actual demolition project starts.

3.2 Health monitoring

Medbioextrem Federal Agency bodies conducted medical sanitary examinations including medical environmental investigation to assess the contamination with super toxic compounds and other hazardous substances in the areas adjacent to JSC Khimprom for about 20 years (1973-1991). These examinations were a part of the sanitary and medical program to ensure the safety of the population during the chemical weapons production.

The concept of safety system includes a set of technological, hygienic and environmental monitoring activities. The basic criterion of environmental acceptability of the former CWPF liquidation is the complete preservation of the health of the population living around the former CWPF as well as the development of safe conditions for work and recreation.

The state sanitary supervision of sanitary epidemiological activities aims at establishing safer labor conditions and at preventing occupational and general somatic diseases as well as at preventing and liquidating environmental pollution is organized in accordance with legislative and executive regulations concerning the facilities under the supervision of the Medbioextrem Federal Agency [6, 15-18]. The state sanitary supervision exists in the forms of preventive and on-going supervision conducted by the Center for State Sanitary Supervision № 29 of the Medbioextrem Federal Agency.

3.2.1 Health examinations

Obviously, the safety of the workers is of first priority during the demolition process. Thus, the labor conditions during the demilitarization procedures can be estimated from the point of view of the possible impact from various hazardous factors on the health of the employees. It should initially be emphasized that in addition to medical effects, psychological effects due to the potential risk of VX poisoning also have to be considered. This can lead to a highly stressful working environment. Furthermore, the elevated temperature in some working areas and the necessity to perform complicated functions while using personal protective equipment may reinforce these effects.

In connection with the medical support during the demilitarization operations, the possibility of both acute and chronic effects on the health of the personnel is considered. However, toxic impact during the regular operations seems to be highly unlikely as only residual amounts of VX remain in the facility today and protective equipment will be properly used during working operations.

The Medical Sanitary Unit 29 (hereafter referred to as Hospital 29), belonging to the Federal Agency "Medbioextrem" of the Russian Health Ministry, conducts the continuous monitoring of the health of the personnel of the former CWPF. The monitoring is conducted in accordance with the orders, medical recommendations and instructions from Medbioextrem [6].

The structure of the former CWPF medical service can be represented as a three-level system:

1. medical sanitary point in building 353, Shop 83
2. policlinic and clinic of Hospital 29
3. specialized clinic of the research institute of occupational hygiene and human environment of Medbioextrem (the Federal Occupational Medicine Center)

The medical sanitary point in building 352 (first level) will perform monitoring of the health of the employees through medical examination before and after the single work shifts. Pre-shift medical examinations are due to the necessity of using personal protection equipment including gas mask and protective suit, which can constitute a strain on the cardiovascular -, the respiratory -, the gastrointestinal – and the urinary organs and the skin.

These daily examinations are based on the observation of a number of health parameters before and after shift. Medical assistants conduct this examination. Since the time allocated for the examination is limited, the establishment of a standard examination program and clear requirements as to the procedures are necessary.

The obligatory procedures are: interviewing, checking the skin, measuring pulse rate, arterial blood pressure, temperature, weight and pupillary reaction testing.

The medical assistant logs all information obtained into the workers' personal medical monitoring card. The entry and exit time is noted and a permission to start work is either issued or denied. Every time an employee is not allowed into the "dirty" working zone, a note is made in a special logbook, which is handed to the foreman of the shift.

The medical assistant at the medical point keeps a strict record of the examined employees. During the after-shift examination signs of toxic impact and possible effects of physical and psycho-physiological laboratory work conditions are assessed. If medical assistance is required, the patient is taken to the polyclinic of Hospital 29.

The medical point is prepared and ready for emergency help in case of an emergency situation. The tasks of the medical point, its equipment and organization are stated in the relevant documents (Medbioextrem Statutory Order no 33 on "Organisation of medical aid to the employees during organo-phosphorus chemical agent destruction").

The second link in the medical support system is the polyclinic and clinic of Medical Sanitary Unit (Hospital) 29. Preliminary and periodic medical examinations are made in the polyclinic once a year, according to Statutory Order No. 101 on "The order of conducting preliminary and periodical medical examinations of the employees engaged in chemical weapons' destruction operations" [19]. Preliminary and periodic medicals are mandatory for all employees whose work is connected with handling toxic substances [20].

The preliminary medical examination includes an examination by specialists such as: internist, surgeon, neurologist, ophthalmologist, otorhinolaryngologist, psychiatrist, dermatologist, and dentist. All employees have taken a hematology and urine test, cholinesterase activity and liver function in the blood, and spirography and electrocardiogram. Other tests, which can be performed, are electromyography and electroencephalography.

The following equipment, which will be provided through the TACIS project, is required to perform these tests: automatic biochemical analyzer, spirometer, electromyograph, electroencephalograph and electric balance. The use of high quality and - performance equipment will improve the health monitoring activities of present and former employees of Plant №4.

The procedures of the yearly periodic medical examination are the same as those for the preliminary examination. The responsible physician keeps a list of all employees who are authorized to work at the facility. This list is forwarded to the medical point.

In case of acute poisoning or aggravation of a chronic disease during working operations, the employees are treated in the specialized treatment unit of occupational safety at Hospital 29.

The third link of medical support system is the specialized clinic of the research institute of occupational health and human environment or the occupational health center of Medbioextrem. In this institution regular medical examinations are performed every three years for all employees. The existing regulating documents also stipulate:

- semi-annual diagnostic follow-ups for employees that have displayed acute or chronic poisoning due to organo-phosphorous compounds.
- life-long follow-ups of former employees that have retired after chemical weapons destruction operations.

In addition to medical assistance, the research institute provides consultations, examines occupational diseases and handles employment problems for the personnel. Further, the institute participates in setting up medical equipment and monitors labor conditions and health of the personnel.

3.2.2 Occupational hygiene and safety monitoring

The creation of safe labor conditions is secured by different means comprising [6]:

- establishment of labor conditions in accordance with the requirements of sanitary regulations and standards, hygienic norms and safety standards and specific requirements regulating sanitary and epidemiological labor conditions
- identification of real risk factors in the occupational environment as well as their sources
- qualitative as well as quantitative assessment of possible effects of single risk factors on the employees, primarily the potential risk arising from residual VX
- creation of risk groups, i.e., grouping of the employees according to potential risks associated with their actual duties

- Continuous monitoring of the personnel health changes in each risk group and verifying possible cause-effect relations between labor tasks and observed health changes.

In connection with ensuring safe working conditions during the liquidation of the former CWPF the experience from the earlier comprehensive sanitary hygienic support during the VX production should be given proper consideration.

The sanitary hygienic system set up during the former CWPF production, includes a) a set of sanitary toxicological requirements including a maximum allowable level of VX content in different media of the environment and b) a set of sanitary hygienic and physiological requirements including such elements as examination of all technical working procedures, development of safe labor conditions including the requirements for optimal protection and physiological acceptability of the personal protection equipment and the development of occupational safety management plans.

Occupational and labor safety instructions have been developed and approved by the Center for State Sanitary Supervision № 29 including requirements to safe operation performance, occupational environment monitoring, the use of personal protection equipment, decontamination, ventilation, sanitary check-point schedule, etc. in accordance with the recommendations on developing rules and instructions on occupational safety and the recommendations of the sate sanitary supervision.

The organization and monitoring of safe labor conditions during the liquidation operations is the responsibility of the occupational safety and industrial control department at Plant № 4.

The toxicological requirements are regarded as fundamental for the monitoring system. Accordingly, a set of VX hygienic standards has been developed for different media (Table 1).

Table 1: Hygienic standards for VX contents in various media.

Medium	***Occupational area air*** [21]	***Surface water*** [6b]	***Soil*** [6b]	***Atmospheric air*** [22]	***Equipment surfaces*** [23]
Unit	mg/m^3	mg/dm^3	mg/kg	mg/m^3	mg/dm^2
Measure[1]	MAC	MAC	PAC	PNeL	MAL
VX	$5x10^{-6}$	$2x10^{-6}$	$1x10^{-5}$	$5x10^{-8}$	$2x10^{-6}$

[1]MAC: Maximum Allowed Concentration; PAC: Predicted Allowable Concentration; PNeL: Predicted No-effect Level; MAL: Maximum Allowed Level

Additionally, emergency impact limits (EIL) have been established for atmospheric air in residential areas at different time intervals (Table 2).

Table 2: Emergency Impact Limits (EIL) for VX contents in atmospheric air [24].

Time hours	***EIL*** mg/m^3
1	1.6×10^{-5}
4	4.1×10^{-6}
8	2.0×10^{-6}
24	6.6×10^{-7}

The monitoring of a safe labor environment is conducted by the laboratories of Shop 83 at Detached Plant № 4 and of the Center for State Sanitary Supervision № 29 and ensures both a quantitative as well as a qualitative assessment of the possible impact of risk factors, basically originating from residual VX, affecting the employees by means of properly established and authorized instrumental and laboratory monitoring methods (*vide supra*).

3.2.3 Industrial monitoring

The 'Federal Act on Sanitary and Epidemiological Health Criteria' [6a] regulates the sanitary and epidemiological health criteria requiring two independent and interconnected monitoring systems. The primary monitoring type is industrial monitoring including laboratory analyses and tests and preventive measures regarding industrial activities.

Industrial monitoring according to the hygienic and sanitary rules and regulations is conducted by the enterprise in accordance with existing standards and regulations [25].

The second system of industrial monitoring is conducted within the framework of the State Sanitary Supervision and is regulated by federal laws and government decrees [6a, 15, and 16]. The State Sanitary Supervision controls the completeness and the quality of the industrial monitoring as an integral part of the preventive measures. The following procedures are used for this activity:

- Environmental monitoring methods in accordance with the activity performed.
- Laboratory testing and research.
- Enforcement of existing regulations and orders.
- Conducting medical examinations.
- Conclusions about the safety for humans and the environment of new types of products, industrial technologies, safety criteria and/or the safety of the industrial and natural environment, as well as the development of monitoring methods and procedures.
- Reporting and accounting of the issues related to the industrial monitoring.
- Timely informing of the population and local authorities and the Center for State Sanitary Supervision № 29 about emergency situations, stoppages and violations of technological processes creating a hazard to public hygiene.
- Inspection by specially appointed people of the observance of the sanitary epidemiological measures and regulation, developing and implementing the measures aimed at removing the identified violations.

The industrial monitoring controls the industrial rooms, equipment, transport, technological processes, workplaces, raw materials, personal protective equipment and production and consumption waste.

3.2.4 Socio-hygienic monitoring

The major component of the existing monitoring system and of the monitoring system to be in place during the demolition operations is socio-hygienic monitoring (SHM). SHM is a state system of observation, analysis, assessment and forecast of the health and living conditions of the population as well as of identification of cause and effect relationships between public health and the impact of the environmental factors [6a].

SHM is conducted by the agencies and bodies of the state sanitary epidemiological service of the Russian Federation together with the executive organs and local authorities. Within the framework of the demilitarization of the former CWPF, detached Plant № 4 at JSC Khimprom, Novocheboksarsk the Center for State Sanitary Supervision of Novocheboksarsk and the Center for State Sanitary Supervision № 29 conduct the SHM on behalf of the state sanitary epidemiological bodies.

During the demolition of the former CWPF automated information analytical system of integrated special socio-hygienic monitoring (socio-hygienic and sanitary-environmental) of changes in the health of the personnel, population and hazardous factors of the single environmental elements will be operating. The system is further designed to disclose major risks and protection both on site and in the residential areas.

The automated information analytical system of special SHM will operate on the basis of two modules, i.e., a national module, operated by NIIGPiECh of Medbioextrem, the Russian Health Ministry, St. Petersburg, and a regional module operated by Hospital 29 of the Federal Agency, the Russian Health Ministry, Novocheboksarsk

According to the Environmental Impact Assessment [11] the demolition process will not alter the overall risk profile of the sanitary medical situation that was rated at ranks 2 or 3, i.e., from relatively to significantly tense. Thus, the health monitoring of the population and socio-hygienic monitoring on the territory of Novocheboksarsk in connection with the demilitarization and demolition of the former CWPF will remain unchanged and will be conducted by the territorial health bodies and the Center for State Sanitary Supervision of the Chuvash Republic.

4. DISCUSSION

Monitoring of hazardous substances possibly being released during the planned demolition of former CWPFs is obviously of utmost importance in order to secure the environment as well as human health. Further, it is crucial to note that health monitoring is strongly dependent on the data obtained through the environmental monitoring activities. The subsequent publication of monitoring data through the proper information channels is necessary to maintain public confidence in and acceptance of the project.

The main focus of the Tacis project is to assist the involved organizations in improving the existing system of environmental and safety and health monitoring of the personnel providing technical assistance.

In relation to the monitoring activities, the main achievements of the Tacis project are expected to be 1) the provision of selected laboratories with equipment for them to comply with the analytical requirements for environmental and health monitoring as foreseen during the demolition of the former CWPF detached Plant № 4 at JSC Khimprom, Novocheboksarsk, Russia, 2) an improvement of the present monitoring systems to assure maximum protection of the workers and the population as well as of the territories surrounding Plant № 4 and 3) to improve the integrated system involving both Health and Environmental Monitoring by adding a system for information dissemination in order to inform in the most optimal way the workers, the media and the population on the current status of the potential environmental and health impact of the demolition work.

Concerning the laboratories and the equipment to be installed, it is important to face the needs for analysis of rather complex environmental samples from a complicated environment. Thus, specific and precise analytical methodology needs to be introduced. The method applied up until now has been based on choline esterase activity assays. This method is unspecific, as it will detect all activity inhibiting choline esterase activity. Thus, both possible VX and "VX equivalents" such as e.g. organo-phosphorus insecticides will be detected. The obvious choice to improve the monitoring activities has been modern gas chromatographic equipment for chemical analyses with specific detectors, the latter being selected according to the actual nature of the compounds to be analyzed, i.e. phosphorous and sulfur selective detectors and mass spectrometric detection. Further, it has been a main consideration that the selected equipment would exhibit a more versatile applicability, i.e. for uses following the completion of the demolition project.

The integration of environmental and health monitoring has recently been emphasized by the European Commission through "A European Environment and Health Strategy" [26]. Thus, as stated, "the interaction between environment and health is far more intimate and complex than is commonly understood" [26].

The destruction of chemical weapons and the demolition of former CWPFs and their waste disposal is a complex chemo-technological process. *A priori* such activities cannot be recognized as safe.

To find solutions to the problem of protecting and improving the human environment, and, consequently, improving the living conditions and public health, it is of crucial importance to have valid and reliable information on the degree of the environmental pollution, on the impact of the environment on public health, on the changes in that impact and its consequences. This raises an urgent question problem of developing an integrated environmental hygienic monitoring, i.e. a system that can provide a comprehensive assessment of environmental changes with regard their negative impact on human health and that can forecast the development of the environmental situation and the changes in public health, thus finding the causes that brought about the negative consequences.

In the European Environmental and Health Strategy [26], the development of a system "integrating information on the state of the environment, the ecosystem and human health" is strongly emphasized. Within the Tacis project, a strong focus is on a close collaboration between the SHM activities and the information activities foreseen to be operated by the newly set-up information center in the City Administration of Novocheboksarsk.

The information activities are a separate important component of the Tacis project. However, it is outside the scope of the present paper to discuss this component further.

5. CONCLUSIONS

The projected demilitarization of the former chemical weapons production facility, the detached Plant № 4 at JSC Khimprom, Novocheboksarsk, Russia, will be accompanied by comprehensive monitoring activities in order to secure minimum impact on the surrounding environment as well as on the health of the workers and the population of the region. A comprehensive environmental impact assessment disclosed the major risk factors for the workers, the population, as well as for the environment.

The EIA concludes that if the accepted solutions are followed and observed, the demolition of the former CWPF, detached Plant № 4, at Khimprom, Novocheboksarsk, will not significantly affect the environment and the living conditions of the population of Novocheboksarsk and the surrounding territories.

An important part of the Tacis project is to supply the involved laboratories with appropriate analytical equipment in order to provide high quality quantitative chemical analyses of VX. Previously applied methods based on the choline esterase assays will be substituted by chromatographic methods using various specific detectors to ensure high sensitivity of the analyses with maximum validity and selectivity. Parallel to the implementation the analytical techniques, appropriate training of the laboratory staff will take place as an integrated part of the improvements of the existing environmental monitoring system.

A close collaboration between all the components of the integrated socio-hygienic monitoring system at all levels, with information exchange at the local level being the crucial issue, are strongly emphasized for providing timely, valid and reliable information for federal and regional government bodies and the population about the sanitary and environmental situation and the measures being taken by the state sanitary supervision agencies to ensure a stable sanitary epidemiological situation on the project territory.

ACKNOWLEDGMENTS

The authors are grateful for financial support from the European Union Tacis program (ENVRUS 9903: Chemical Weapons III).

References

[1] Convention on the Prohibition of the Development, Production, Stockpiling and the Use of Chemical Weapons and on Their Destruction // Конвенция о запрещении разработки, производства, хранения и использования химического оружия и его уничтожении. http://www.opcw.org/html/db/cwc/eng/cwc_frameset.html

[2] a) News Chronology, "24 April", The CBW Conventions Bulletin, no. 60, June 2003, p, 48; b) Federal Law 138, Nov. 5th, 1997 "On the Ratification of the Convention on the prohibition of the development, production, stockpiling and use of chemical weapons" // Закон 138-ФЗ «О ратификации конвенции о запрещении разработки, производства, накопления и применения химического оружия и его уничтожении» от 5.11.1997.

[3] Resolution of the Government of the Russian Federation, № 510, July 5th, 2001 "On making amendments and additions to the resolution by the Government of the Russian Federation of March 21, 1996 (№ 305) on approving the special federal program 'Chemical weapons stockpiles destruction in the Russian Federation'" // Федеральная целевая программа «Уничтожение запасов химического оружия Российской Федерации». Постановление правительства РФ от 21.03.1996 № 305 (в редакции постановления правительства РФ от 05.07.2001 № 510).

[4] a) "Detailed plans for destruction of the facility for the production of VX-type substance and filling it into munitions, Open Joint Stock Company "Khimprom", Novocheboksarsk, Chuvash Republic, Russian Federation (Phase 1), OPCW Executive Council, EC-XX/NAT.4, 4. May, 2000 // Подробный план уничтожения объекта по производству вещества типа Vx и снаряжения им боеприпасов, открытое акционерное общество «Химпром», Новочебоксарск, Чувашская Республика, Российская Федерация (1 этап), исполнительный комитет ОЗХО, EC-XX/NAT.4, 4 мая 2000 г.; b) "Request for conversion and detailed plans for the facility for production of VX type substances and filling it into munitions (Stage II: Conversion of part of the auxiliary buildings 352 and 353 and of the ventilation stack 366B) at JSC "Khimprom", Novocheboksarsk, Russian Federation" OPCW Executive Council, EC-XXVII/NAT.2, 17. August 2001 // Запрос на конверсию и подробные планы объекта по производству вещества типа Vx и снаряжения им боеприпасов (этап 2: конверсия части вспомогательных корпусов 352 и 353, вентиляционной трубы

366Б) на ОАО «Химпром», Новочебоксарск, Российская Федерация, исполком ОЗХО EC-XXVII/NAT.2, 17 августа 2001г.

[5] Organisation for the Prohibition of Chemicals Weapons, http://www.opcw.org. // Организация по запрещению химического оружия.

[6] a) Federal Law 52 of 30.09.99 "On Sanitary Epidemiological health criteria" // Закон от 30.09.99г. № 52-ФЗ "О санитарно-эпидемиологическом благополучии населения"); b) Instructions 1.1.019-00 "Organization and provision of sanitary epidemiological supervision at the organo-phosphorus chemical agents destruction facilities" // МУ 1.1.019-00 "Организация и осуществление санитарно-эпидемиологического надзора на объектах по уничтожению фосфороорганических отравляющих веществ"); c) Instructions 2.2.21016-98 "Labor Conditions Safety during Operations with Chemical Agents in the Laboratories and Experimental Units of the Research Institutes, agencies and organizations" // МУ 2.2.1.016-98 "Безопасность условий труда при работе с отравляющими веществами в лабораториях и экспериментальных подразделениях научно-исследовательских институтов, учреждений и организаций"; d) Instructions 2.2.508-02 "Setting up of Sanitary Chemical Control over the Occupational Environment Condition at the organo-phosphorus Chemical Agents Destruction Facilities" // МУ 2.2.5.08-02 "Организация санитарно-химического контроля за состоянием производственной среды на объектах по уничтожению фосфорорганических отравляющих веществ"; e) Statutory Order 101 of 21.03.00 of the Russian Health Ministry "On the order of conducting preliminary and periodic medical examinations of the employees engaged in chemical weapons destruction operations" (Приказ Минздрава Российской Федерации № 101 от 21.03.00 г. "О порядке проведения предварительных и периодических медицинских осмотров работников, занятых на работах по уничтожению химического оружия"; f) Instructions on conducting preliminary (upon hiring) and periodic medical examinations of the employees engaged in organo-phosphorus chemical agents destruction, approved 26.06.00 (Методические рекомендации по проведению предварительных при поступлении на работу и периодических медицинских осмотров работников, занятых на работах по уничтожению ФОВ, утв. 26.06.00 g) Statutory order of the Medbioekstrem Federal Agency, Russian Health Ministry № 33z of 28.06.99 "On establishing medical aid to the personnel of the organo-phosphorus chemical agent destruction facilities" (Приказ Федерального Управления «Медбиоэкстрем» при Минздраве России № 33з от 28.06.99 г. "Об организации медицинской помощи работникам объектов по уничтожению ФОВ". h) Statute on an industrial sanitary laboratory of the state sanitary supervision centres, Medical-biological and emergency problems agency, Russian Health Ministry, supervising chemical weapons destruction facilities, approved 18/06/2002 // Положение о промышленно-санитарной лаборатории центров Госсанэпиднадзора федерального Управления медико-биологических и экстремальных проблем при Министерстве здравоохранения Российской Федерации, обслуживающих объекты уничтожения химического оружия, утв. 18.06.02.

[7] Federal Law No 76 of 02/05/1997 On Chemical Weapons Destruction. // Федеральный закон «Об уничтожении химического оружия» от 02.05.97 № 76-ФЗ

[8] Federal Law No 7 of 10/01/2002 On Environmental Protection. // Федеральный закон «Об охране окружающей среды» от 10.1.02 № 7-ФЗ

[9] Federal Law No 08 of 23/12/1994 On the protection of the population and the territories from disasters and technogenic emergencies. // Федеральный закон «О защите

населения и территорий от чрезвычайных ситуаций природного и техногенного характера» от 23.12.1994 № 8-ФЗ.

[10] Terms of Reference for the Tacis Chemical Weapons III project ENVRUS 9903/ Техническое задание к проекту Тасис «Химоружие 3» ENVRUS 9903.

[11] "Environmental impact assessment (EIA) of the project "Liquidation of the consequences of the former chemical weapons production facilities at Detached Plant № 4 of JSC Khimprom, Novocheboksarsk, Chuvash Republic in connection with the 'Convention on the Prohibition of the Development, Production, Stockpiling and the Use of Chemical Weapons and on Their Destruction'" // Оценка воздействия на окружающую среду (ОВОС) к проекту «Ликвидация последствий деятельности бывших объектов по производству химического оружия на Обособленном заводе №4 ОАО «Химпром» г.Новочебоксарск Чувашской Республики в соответствии с Конвенцией по запрещению разработки, производства, хранения и использованию химического оружия и его уничтожения.

[12] "Chemical weapons – threats, effects and protection" (2002), FOI, Swedish Defence Research Academy, briefing Book, No 2

[13] P.Ye. Shkodich. Scientific foundations of the sanitary hygienic support of chemical weapons destruction. In: Occupational Medicine and Industrial Ecology. 12. 2002. // П.Е. Шкодич. Научные основы санитарно-гигенического обеспечения процесса уничтожения химического оружия. // Медицина труда и промышленная экология. - № 12. - 2002.

[14] Industrial Monitoring Programme over sanitary rules observance and sanitary epidemiological (preventive) measures at Detached Plant № 4 and at JSC Khimprom. // Программа производственного контроля за соблюдением санитарных правил и выполнением санитарно-противоэпидемических (профилактических) мероприятий по обособленному заводу № 4 и в целом ОАО «Химпром».

[15] Federal Law No 134-FL of 08.08.2001 On Protecting the rights of legal entities and individual entrepreneurs during the state supervision // Федеральный закон «О защите прав юридических лиц и индивидуальных предпринимателей при проведении государственного контроля (надзора)» от 08.08.2001 №134-ФЗ.

[16] Resolution of the Government of the Russian Federation No 554 of 24.07.2000 On approval of the Statute of the state sanitary epidemiological service of the Russian Federation and the Statute of the State sanitary epidemiological regulation // Постановление правительства Российской Федерации от 24.07.2000 №554 «Об утверждении Положения о государственно-эпидемиологической лужбе Российской Федерации и Положения о государственном санитарно- эпидемиологическом нормировании».

[17] Statutory Order of the Russian Health Ministry No 76 of 19/03/2002 On approval of the Regulations of the procedures of the sanitary epidemiological supervision at the facilities and on the territories supervised by the Medbioextrem Federal Agency // Об утверждении Положения о порядке осуществления государственного санитарно-эпидемиологического надзора на объектах и территориях, обслуживаемых федеральным управлением «Медбиоэкстрем». Приказ Минздрава РФ от 19.03.2001 № 76.

[18] Statutory Order of the Russian Health Ministry No 228 of 17/072002 On the procedures of conducting monitoring activities during state sanitary epidemiological supervision // Приказ Минздрава РФ от 17.07.2002 № 228 «О порядке проведения мероприятий по контролю при осуществлении государственного санитарноэпидемиологического надзора».

[19] Order of statute of the Federal Medbioextrem Agency № 33a, June, 28th, 1999 "On organizing medical assistance to the personnel of the organo-phosphorus compound destruction facilities" // Приказ ФУ Медбиоэкстрем №33з от 28.06.1999 «Об организации медицинской помощи работникам объектов по уничтожению ФОВ».
[20] Resolution of the Government of the Russian Federation No 188 of 29/03/2002 On approval of the lists of productions, professions and occupations with hazardous labour conditions which make those employed for work with chemical weapons eligible for benefits and compensations // Постановление правительства Российской Федерации от 29.03.2002 г. №188 «Об утверждении списков производств, профессий и должностей с вредными условиями труда, работа в которых дает право гражданам, занятых на работах с химическим оружием, на льготы и компенсации».
[21] Hygienic norms of maximum allowable concentrations (MAC) of occupational air in the chemical weapons storage and destruction facilities HN 2.2.5.1371-03 // ГН 2.2.5.1371-03 Гигиенические нормативы предельно-допустимых концентраций (ПДК) вредных веществ в воздухе рабочей зоны объектов хранения и уничтожения химического оружия».
[22] Hygienic norms of predicted no-effect impact level (PNeL)of contaminants in the atmospheric air of residential areas and in the protection activity zones of the chemical weapons storage and destruction facilities HN 2.1.6.1372-03 // ГН 2.1.6.1372-03 «Гигиенические нормативы ориентировочных безопасных уровней воздействия (ОБУВ) загрязняющих веществ в атмосферном воздухе населенных мест и зонах защитных мероприятий объектов хранения и уничтожения химического оружия».
[23] Maximum allowable level (MAL) of the technological equipment contamination (with coating and without it), appr. 24.07.1996 No 08-86. // Предельно-допустимый уровень загрязнения технологического оборудования (с покрытием и без покрытия), утв.24.07.96 №08-96.
[24] Hygienic norms Emergency impact levels (EIL) of chemical agents for atmospheric air in the residential areas HN 2.1.6.1181-02. // ГН 2.1.6.1181-02 «Аварийные пределы воздействия (АПВ) отравляющих веществ для атмосферного воздуха населенных мест».
[25] SP 1.1.1058-01 Organization and procedures industrial monitoring of sanitary rules observance and sanitary epidemiological (preventive) measures. // СП 1.1.1058-01 «Организация и проведение производственного контроля за соблюдением санитарных правил и выполнением санитарно-противоэпидемических (профилактических) мероприятий.
[26] European Commission (2003). "A European Environment and Health Strategy", Communication from the Commission to the Council, the European Parliament and the European Economic and Social Committee, COM (2003) 338 final.

VAPOR VALIDATION OF MONITORING SYSTEMS FOR DETECTION OF TRACE LEVELS OF CHEMICAL WARFARE AGENTS IN AIR

Joseph Padayhag[1]
[1]*Department of Health and Human Services, Centers for Disease Control and Prevention (CDC), Atlanta, Georgia (U.S.A.)*

Abstract: U.S. Army facilities for elimination of chemical warfare agents (CWA) include instrumentation for monitoring trace-levels of CWA in air. Protocols for calibration and quality control assessment of air-monitoring systems use solutions of agent standards in solvent; however, agent potentially detected in the facility would be in gaseous form. The Army is conducting vapor validation studies in collaboration with the Centers for Disease Control (CDC)/National Institute for Occupational Safety and Health (NIOSH). These studies assess performance of instrumentation to monitor for CWA vapor at temperature and humidity conditions encountered at CWA destruction facilities. Vapor validation studies on monitoring methods for lewisite (L) [dichloro-(2-chlorovinyl)arsine] are of interest because the CWA is difficult to sample and must be derivitized for analysis. Vapor validation efforts have resulted in recommendations for changes to equipment configuration to enhance monitoring of CWA vapor.

Keywords: vapor validation, lewisite

1. INTRODUCTION

U.S. Army facilities and systems for chemical warfare agents' (CWA) destruction include instrumentation for monitoring trace levels of CWA in air. Protocols for calibration and quality control assessment of monitoring systems in CWA destruction facilities use solutions of agent standards in solvent because of difficulties in generating CWA at quantifiable low levels in air.

The Army is conducting vapor validation testing based on CDC/NIOSH protocol to demonstrate that liquid spiking procedure is valid for calibration, quality control evaluation, and monitoring CWA in a gaseous stream.[2] The sampling of a generated atmosphere with candidate and independent methods allow determination of (a) the capacity of the sampler, (b) the efficiency of analyte collection by the sampler (c) the repeatability of the

V.M. Kolodkin and W. Ruck (eds.),
Ecological Risks Associated with the Destruction of Chemical Weapons, 199–203.

method, and (d) the bias of the method, and (e) interferences in the collection of the sample. This paper will focus on recent vapor validation studies for L.

2. EVALUATION PROTOCOL

NIOSH evaluation protocol guidelines include [1]:

a) Selection of an independent comparison method.
b) Evaluation of candidate methods at multiple concentration levels. For L, concentrations were chosen at 0.8 and 1.8 times the time-weighed average (TWA) workplace exposure limit. The L TWA is equal to 3×10^{-3} mg/m^3.
c) Evaluation at appropriate environmental parameters. Five environmental conditions were selected on the basis of temperature and humidity conditions of CWA disposal facilities listed as follows:
   1. 60°F (16°C) and 22% relative humidity
   2. 75°F (24°C) and 35% relative humidity
   3. 75°F (24°C) and 50% relative humidity
   4. 75°F (24°C) and 65% relative humidity
   5. 90°F (32°C) and 90% relative humidity

Data collected from candidate methods were compared with the independent reference method. Data were then analyzed to determine method bias, precision, and accuracy. Ideally, methods should meet criteria of plus or minus (+/-) 25% accuracy with 95% confidence. If methods fail to meet the accuracy criteria, they may continue to be viable if operational conditions are implemented, such as lowering the alarm setpoint.

3. LEWISITE MONITORING METHODS

Two types of monitoring systems are (a) near-real time (NRT) monitors, comprising both sampler and analyzer, to provide early warning of potential worker exposure and process-upset conditions in 15 minutes or less and (b) NRT alarm confirmation/historical monitors, which comprise sample collectors analyzed in a laboratory on site. The vapor validation studies included one NRT candidate method, two alarm confirmation/historical methods, and a reference method.[3, 4] In this paper, the term "method" refers to the instrument tested, but is meant to include procedures and parameters for sampling and analyzing.

Equipment for the NRT candidate method was a MINICAMS®, manufactured by O.I. Analytical (Birmingham, AL), configured to monitor for L. The sample stream was derivatized using ethanedithiol gas and transported through a heated Teflon® sample line to the MINICAMS®. The MINICAMS® included a small sorbent tube to concentrate the sample, a capillary gas chromatography (GC) column, and a halogen selective detector (MINICAMS®-XSD).

The two NRT alarm confirmation/historical candidate methods are described as follows:

a) An ascorbic acid bubbler, where L was collected, converted to acetylene, and analyzed with a GC configured with flame ionization detector (FID).
b) A depot-area agent-monitoring system (DAAMS), where CWA was collected on a Tenax TA® sorbent tube. For this method, the sample was extracted with hexane and methanol. The extract was spiked with hydrogen peroxide to convert both L and chlorovinylarsenous acid, a hydrolysis product of L, to chlorovinylarsonic acid. The spiked extract was analyzed using high-performance liquid chromatography with mass spectrometry (MS) detection.

The reference method was a nitric acid bubbler analyzed for arsenic by graphite furnace/atomic absorption based on Environmental Protection Agency method 7060A.

4. VAPOR VALIDATION RESULTS

Results are summarized as follows:[3, 4]

a) The MINICAMS® method demonstrated +/- 42% accuracy at a 95% confidence interval and exhibited a positive bias of vapor versus liquid challenges at most test conditions, which was interpreted as conservative for worker protection.
b) The DAAMS methods demonstrated +/- 35% accuracy at a 95% confidence interval.
c) The ascorbic acid/GC-FID method proved unable to recover L vapor during the validation testing, and validation testing was discontinued.

The Army and CDC concurred that the MINICAMS® and DAAMS methods could be used for L destruction facilities at normal temperatures/relative humidity with the alarm set point reduced to 40% of the L-TWA.

The testing resulted in several recommendations for the MINICAMS®-XSD as a result including eliminating restrictions to the probe before derivitizing, diverting flow from the probe filter during non sampling intervals, and replacing flow meters with flow controllers.

5. CONCLUSIONS

The MINICAMS® real time method and DAAMS confirmation methods are valid for monitoring 1 TWA of L in a gaseous stream at environmental conditions of CWA disposal facilities at typical environmental conditions with appropriate adjustment in alarm set point and increased quality control measures. Furthermore, liquid challenging is an acceptable means of verifying the quality of L MINICAMS® and DAAMS monitoring systems.

Even though liquid challenging is an acceptable means of conducting quality control challenges, it does not reflect the true accuracy of the instrument. The MINICAMS® method response varied more with vapor challenge than with liquid injection. The ascorbic acid bubbler could detect L in liquid solution and demonstrated +/-25% recovery to liquid challenges, yet was incapable of recovering L from a vapor stream.

Several modifications and recommendations for changes in instrument configuration have resulted from vapor validation studies. Restrictions to the sample probe before derivitization and areas where CWA may condense are not problematic for liquid injection, but interfere with vapor sampling. The use of a flow meter rather than a flow controller is not problematic for analyzing small volumes of CWA in liquid solvent; however, flow rate becomes critical for determining sample volume of CWA vapor.

References

[1] U.S. Army (2001). Programmatic Laboratory Quality Assurance Plan, Final Revision 2, U.S. Army Program Manager for Chemical Demilitarization, Aberdeen Proving Ground, MD.

[2] National Institute of Occupational Safety and Health (NIOSH) (1995). Guidelines for Air Sampling and Analytical Method Development and Evaluation, NIOSH, Cincinnati, OH.

[3] U.S. Army (2002). Vapor Validation Test of Lewisite MINICAMS®, Bubbler with GC/FID, and DAAMS with HPLC/MS. Draft Report. U.S. Army Program Manager for Chemical Demilitarization, Aberdeen Proving Ground, MD.

[4] U.S. Army (2003). Lewisite Vapor Validation for the MINICAMS®-XSD and DAAMS Methods. Final Report. U.S. Army Program Manager for Chemical Demilitarization, Aberdeen Proving Ground, MD.

TACIS PROJECT
"THE DEVELOPMENT OF AN ENVIRONMENTAL AND HEALTH MONITORING SYSTEM RELATED TO THE DEMILITARIZATION OF THE CHEMICAL WEAPONS FACILITY AT THE DETACHED PLANT NR. 4 OF OAO KHIMPROM, NOVOCHEBOKSARSK"
AND POSSIBLE IMPLICATIONS FOR FUTURE RUSSIAN DEMILITARIZATION ACTIVITIES AT CWPFS

Thomas Stock

*Tacis Chemical Weapons III Project, Novocheboksarsk (Russia)**

Abstract: The Tacis Chemical Weapons III is focused upon support of the development of an environment and health monitoring system related to the demilitarization activities at the former Chemical Weapons Production Facility, the detached plant no. 4 at JSC Khimprom, Novocheboksarsk, Russia. The project aims, beside the delivery of equipment for improving the existing monitoring capabilities, at supporting institutional building and co-operation between Russian agencies, organizations and institutions involved in monitoring activities.

Key words: Tacis, former Chemical Weapons Production Facility (CWPF); demolition; safety, health and environmental monitoring;

1. INTRODUCTION

The Russian Federation has ratified the Chemical Weapons Convention in the fall of 1997.[1] The former Soviet chemical weapons (CW) complex

* For correspondence: DYNASAFE Germany GmbH, Düsseldorfer Straße 138, 45481 Mülheim an der Ruhr, Germany, e-mail: thomas.stock@dynasafe.de

V.M. Kolodkin and W. Ruck (eds.),
Ecological Risks Associated with the Destruction of Chemical Weapons, 205–210.

was quite formidable. Declared at 40,000 metric tons and stored at seven sites, Russia possessed the world's largest chemical arsenal.[2]

In January 1998 the Russian Federation supplied to the Organisation for the Prohibition of Chemical Weapons (OPCW) its initial declaration in which 24 [3] chemical weapons production facilities had been listed. The required initial inspections of all declared facilities, including the 7 storage facilities, were completed by August 1998.[4] Because of the fact that former chemical weapons production facilities (CWPFs) in Russia have been parts of large industrial chemical complexes, Russia has asked for permitting conversion of former CW manufacturing installations, for which the CWC provides possibilities.

In 2000 Russia submitted plans for the destruction and verification of the large VX production and filling facility at Novocheboksarsk, Khimprom attached plan no. 4.[5] The plans contain details concerning the planned destruction of equipments and the building of the former VX CWPF.

2. TACIS PROGRAMME

Launched by the EC in 1991, the Tacis [6] Programme provides grant-financed technical assistance to 13 countries of Eastern Europe and Central Asia, and mainly aims at enhancing the transition process in these countries. Since 1991, Tacis has granted over 2.46 billion euros for the implementation of more than 1,500 projects in the Russian Federation. When Tacis was initiated, technical assistance was a stand-alone activity, whereas the programme is now part of a complex and evolving relationship with each of the 13 countries concerned. With the implementation of the Agreement on Partnership and Cooperation (PCA) between the European Communities, their Member States and the Russian Federation, which entered into force on 1 December 1997, as well as the EU enlargement process, Tacis also becomes a more strategic instrument in the co-operation process between EU and partner countries.

3. TACIS'S SUPPORT OF THE RUSSIAN CW DESTRUCTION PROGRAMME

After Russia had ratified the CWC, the European Council issued a Declaration on Assistance to the Russian Federation related to the Convention, allowing the European Commission to allocate €10 million to Tacis projects in support of implementation. It was to be used for technical assistance for conversion and restructuring of former chemical weapons industries as well as on environment and safety related issues.

Tacis launched two projects under the programme of supporting the Russian Federation in its activities to comply with the obligation under the CWC.[7] The beneficiary is the Russian Munitions Agency (RMA). After tendering the third project "ENVRUS 9903: Chemical weapons III: Development of an environmental, sanitary and health monitoring system in connection with the demilitarisation of the chemical weapons facility at OAO Khimprom, Novocheboksarsk, Chuvash Republic" in 2002, the contract was awarded to a consortium headed by COWI, one of the leading Danish engineering companies, in December 2002. Dynasafe GmbH (Germany), National Environmental research Institute (NERI) (Denmark) and Clinic of Occupational and Environmental Medicine (COEM) (Denmark) are the other members of the Consortium. The project duration is until August 2004.

4. PROJECT FOCUS

The Tacis project "Chemical weapons III" is concerned with plant #4, which produced the VX and filling munitions with VX (buildings 350, 352, 353 and ventilation tubing 366 A and B). This building is now residually contaminated with VX decomposition products and must be completely destroyed.

A destruction scheme, which cost some 40 to 45 million EUROS, and meeting the CWC deadline of 29 April 2007, has been approved. Preparatory actions and demolition work have begun.

The project for the demolition and related works is funded from the Russian budget, with financial support from the US Cooperative Threat Reduction Agreement Programme [8]. The present Tacis project aims to help Russia to fulfil related Convention obligations requiring the development of an environmental and sanitary monitoring system to accompany the works and to serve the chemical plant after this.

The overall project objective is to support a sustainable demilitarisation of the Khimprom facility, building 350 and ventilation tubing 366B aimed at producing VX and filling munitions with VX. To achieve this, the project will build on the experience of the previous Tacis projects in the field in order to: *(1)* Develop and start up sustainable Health and Environmental Monitoring Systems (H&EMS) building on the existing networks and institutions in the region including, but not limited to, Medical and Sanitary Unit MChS-29 and the Sanepidnadzor environmental monitoring unit. *(2)* Provide information to the public on the procedures adopted for safeguarding health and the environment in relation to the demolition and disposal works. *(3)* Undertake relevant training for environmental and health monitoring systems initially related to the chemical weapon destruction programme. *(4)* Work with relevant federal and regional administrations responsible for the funding and management of health and environmental monitoring systems to ensure complementarity of information. *(5)* Ensure dissemination of results.

In accordance with the general aim of the project, its specific objectives are: *(1)* To improve the monitoring system established for the demolition and disposal targeted on the control of air, water and soil quality as well as human health during actual work; *(2)* To evaluate and improve already applied sampling and testing procedures in close cooperation with federal and regional institutions; *(3)* To support the provision of information to the population on safe demilitarization activities; *(4)* To disseminate results related to environmental and health monitoring.

The Russian partners in implementation of the project are: *(1)* Munitions Agency of the Russian Federation; *(2)* Detached plant #4 of OAO Khimprom, Novocheboksarsk; *(3)* Medical and Sanitary Unit MSCh-29 (organ of the Ministry of Health); *(4)* Committee for Natural Resources, Chuvash Republic; and *(5)* Committee for Civil Protection and Emergency Situations, Chuvash Republic.

5. PROJECT ACTIVITIES

To meet the project objectives and provide the expected outputs the Consortium has decided to group the activities in 4 thematic modules and organise the work in 4 groups of experts accordingly. Therefore the Project is structured into 4 core components:

Component 1: Procurement and Project Management

The activities under this component will include contacts to relevant federal and regional administrations responsible for the funding and management of the monitoring systems.

Component 2: Health and Safety Monitoring

The activities under this component are focussed upon an implementation of a reliable monitoring system for health and safety during and after the decommissioning.

Component 3: Environmental Monitoring

This component's activities are focused on reinforcement of the existing environmental monitoring system, including capacity building of the institutions involved.

Component 4: Information Management

Activities under this component will provide the outputs related to analytical centre and information activities, including information to the population on safe demilitarising activities.

The project is divided into 4 phases: *(1)* Inception phase; *(2)* Planning phase; *(3)* Implementation phase and support activities; *(4)* Concluding phase with the final report before the 30th of August 2004. In the course of the project at least two study tours for Russian local experts and beneficiaries are organized. The project has established a project office at Novocheboksarsk. The project is using for its work on-site various local experts. In the course of the project seminars, workshops and press briefings are conducted.

Under the project certain funds are set aside for purchasing equipment for the improvement of the environmental, health and safety monitoring system established in close co-operation with the beneficiaries.

6. CONCLUSIONS

The Tacis Chemical Weapons III project supports the demolition work planned at the former CWPF, the detached plant no. 4 at JSC Khimprom, Novocheboksarsk, Russia. With the delivery of different equipment the monitoring capabilities of the special monitoring institutions and agencies will be strengthened. In addition, the project will support the co-operation between different monitoring institutions and support the establishment of an analytical and information centre through which public will be better informed about the possible risks the demolition work might have.

ACKNOWLEDGMENT

The author is grateful for financial support from the European Union Tacis programme (ENVRUS 9903: Chemical Weapons III).

References

[1] Russia deposited its instrument of ratification on 5 November 1997. See SIPRI Yearbook 1998: Armaments, Disarmament and International Security, p. 458.

[2] IPRI Yearbook 1999: Armaments, Disarmament and International Security, p. 573.

[3] Utkina, S, Grobovsky, A. and Zhuchkov, A., "Russian views on conversion of former chemical weapons production facilities", OPCW Synthesis, no. 5 (Nov./Dec. 1999), pp. 1, 13-14; see also APPENDIX NO. 3 to the Federal purpose oriented Program "Chemical Weapons Stockpiles Destruction in the Russian Federation", The Government of the Russian Federation, Resolution No. 510, of July 5, 2001, Moscow, "On Making Amendments and Additions to the Resolution by the Government, of the Russian Federation of March 21, 1996 (No. 305)", On Approving the Federal Special Program, "Chemical Weapons Stockpiles Destruction in the Russian Federation".

[4] SIPRI Yearbook 1999: Armaments, Disarmament and International Security, p. 569.

[5] OPCW Secretariat Brief, no. 24 (21 December 2000).

[6] Tacis: Technical Assistance for the Commonwealth of Independent States.

[7] One on the development of an environmental monitoring system for the area surrounding the Gorny Chemical Weapons Destruction facility in the Saratov oblast and a second on addressing human safety and environment related issues resulting from the destruction of chemical weapons production facilities at Dzerzhinsk in the Nizhny Novgorod oblast.

[8] Including support in demolition and demilitarisation, including laboratory equipment and testing of processes for contaminated metal degasification.

TOXICOLOGICAL AND PUBLIC HEALTH ASPECTS OF TWO-STAGE TECHNOLOGY OF CHEMICAL WARFARE AGENTS DESTRUCTION IN RUSSIA

Andrey Radilov[1]

[1]*Research Institute of Hygiene, Occupational Pathology and Human Ecology, St.Petersburg (Russia)*

Abstract: The complexity of multi-component mixtures of unknown composition and biological activity that are formed during destruction of chemical weapons poses the problem of their thorough toxicological and public health study. The general procedure scheme of toxicological and health investigations of a complex chemical factor of unknown composition and toxicity vector is suggested, which includes several main blocks: information, sample preparation, analytical, toxico-sanitary, biological, chemical markers, biological markers, and evaluation.

Key words: chemical weapons, bitumen-salt mass, chemical markers

The Russian two-stage technology of chemical weapons destruction involving chemical detoxication of chemical warfare agents (CWA) at the first stage followed by subsequent bituminizing of neutralized products with the aim of producing non-soluble bitumen-salt mass (BSM) requires thorough study because of the high hazard of initial components, inadequate study of CWA destruction products, and a priori their unpredictable composition. The probability of polluting environmental recipients with CWA destruction products from gasous emission, or from leaching by rainfall or ground water with subsequent contamination of drinking water, must not be completely ruled out, either during destruction of chemical weapons, or during BSM long storage or transportation of reaction masses after destruction. The complexity of multi-component mixtures of unknown composition and biological activity that are formed during destruction of chemical weapons poses the problem of their thorough toxicological and public health study. The complexity of the problem required formation of a concept of toxico-sanitary evaluation of a complex chemical substance with unpredictable biological effects and composition, as well as to develop methods of solving similar problems.

In this report the general procedure scheme of toxicological and health investigations of a complex chemical substance of unknown composition and toxicity vector is suggested. It includes several main blocks:

V.M. Kolodkin and W. Ruck (eds.),
Ecological Risks Associated with the Destruction of Chemical Weapons, 211–215.

information, sample preparation, analytical, toxico-sanitary, biological, chemical markers, biological markers, and evaluation.

The toxico-sanitary block involves traditional approaches of preventive toxicology, data on toxic properties, thresholds of acute and chronic action, mechanism of action, biological transformation, indication, behavior in the environment, as well as validation of sanitary standards for air, water, working zone, and skin and developing therapeutics and preventives. The traditional toxico-sanitary evaluation is a starting point and at the same time an integrated component of the methodology (suggested by us) that has been formulated due to constant necessity of solving complex scientific and practical problems connected with chemical substances' exposure. Yet, evaluation of danger of the multi-component mixtures with previously unpredictable or variable composition is impossible in the case when only the classical approach is used. CWA destruction products fall exactly into this category of substances.

Let us consider more thoroughly the algorithm of toxico-sanitary evaluation of danger by the example of an urgent situational problem – evaluation of hazard of GB, GD, VX, and mustard-lewisite mixtures.

The technology developers indicate logical components (from the standpoint of the chemical aspect) of the reaction of detoxication in the baseline data. For the most part these are low-solubility and low-hazard calcium salts. Toxicometric data support BSM low hazard in the case of its intragastric administration, visible signs of intoxication in the case of saturating concentrations of vapors in air are absent, local and resorption effects are not observed, and the tendency for a cumulative lethal effect is weak. However, toxicity of one of the products of bitumen-salt mass after VX destruction rises sharply in case of intraperitoneal administration, which is accompanied by the clinical picture typical of organophosphorus CWA, including inhibition of cholinesterase in plasma and erythrocytes. During system investigations, as part of the toxicological block, other effects of CWA destruction products were noted, namely: mild embryotoxic effects of BSMs of mustard-lewisite mixtures and organophosphorus CWA, mild gonadotoxic effects on males and weak effects on females, sensitization to organophosphorus compounds revealed by specific cell reactions – RILM (reaction of inhibition of leukocyte migration) and RSAL (reaction of specific agglomeration of leukocytes). Mutagenic effects were not shown by the dominant lethal mutations and micronucleus tests.

In the biological block, when testing in vitro the destruction product of mustard-lewisite mixture in hydrobionts, toxicity on daphnia and paramecium was shown, and in the open enzyme-and-substrate system changing the glycolysis rate and SH-groups content was noted.

The findings obtained allowed threshold and non-effective concentrations of aqueous extracts of organophosphorus CWA destruction products to be established for rats and hydrobionts.

The toxicological block made it possible to reveal biological markers and biological response and to formulate recommendations for health monitoring of persons engaged at chemical weapons destruction facilities and for the population residing in their vicinity, on the basis of the symptoms observed during the experiment.

The next block is preparation of samples and evaluation of probable routes of the delivery of a chemical substance. The main procedure consists of modeling saturating concentrations of substances in contact media for subsequent study in toxico-sanitary, biological and analytical blocks.

In the analytical block the 1st stage includes the prospecting analysis of contact media that are conductors of a hazardous substance. The chromato-mass spectrometry is used to determine the total spectrum of compounds and to choose the most persistent ones in the environment. Mustard markers are thioglycols, 1,4-ditiane, and ethanolthiomorpholines. Markers for each of the media can differ, this difference being caused by physicochemical properties and degradation rates. After determining the list of markers specific methods of analytical monitoring are developed to accomplish the analysis in the environment and monitoring of the delivery of chemical substances into the body.

Among products of GB, GD and VX destruction under conditions of the two-stage technology we have isolated homologous series of mono- and diesters of methylphosphonic acid. These compounds, as well as their hydrolysis product - methylphosphonic acid - have been chosen as chemical markers to evaluate probable pollution of the environment and indicate presence in biological media, as well as to prove the interaction with organophosphorus CWA and their destruction products. The total list of chemical markers in the environment and in biological media that establish the occurrence of exposure to CWA or their destruction products is given in the table below.

Thus, the concept of toxico-sanitary evaluation of an unknown chemical factor (substance) includes four elements: 1) the system of chemical compounds and matrices of complex chemical composition investigation; 2) the main criterion of evidence of the body and xenobiotic interaction; 3) the notion of "xenostatus"; 4) the main object of preventive toxicology.

The system of studying the matrix of a complex chemical composition is based on treating it as a subject with unknown properties or "black box", which allows errors of its evaluation to be avoided. The system includes five main blocks: sample preparation, analytical, toxicological, biological, and appraising.

The sample preparation is carried out to model passing of a chemical substance into contact media through which harmful chemical compounds are delivered into the human body under conditions simulating the real ones (temperature, media composition, pressure, humidity, and other parameters) as much as possible.

The composition and the concentration of migrating substances are determined by physicochemical methods. Persistent substances, whose greatest concerns are toxicity and hazard, are revealed. The destruction process of chemicals is studied with time. Persistent compounds or destruction products that are available in matrices are revealed, and they can serve as markers for environmental pollution for those of the body interaction with xenobiotics in case of their detection in biological media.

Toxicity of separate substances, mixtures, as well as products of migration into biological media, is studied for mammalians and hydrobionts with the aid of methods of traditional toxicology. In case of probable wide spread or long-term action of a harmful substance, delayed effects (embryotoxic, gonadotoxic, mutagenic, immunotoxic and others) are studied. In this case the standard procedure of hygienic rating of a harmful substance is used with development of the monitoring method.

In our concept particular attention is given to monitoring the delivery of a chemical substance into the human body. Solving the problem of the reliable determination of the fact of the interaction with a substance and its quantitative characteristics is considered very urgent. This is made possible in two ways: exposing chemical markers of the interaction (substances, their destruction products, and biological derivatives) in biological media, or revealing biological responses. Yet, the most exact evidence is one obtained by both methods.

Therefore, in addition to the known clinical-and-sanitary and epidemiological group methods of demonstrating the effects of a chemical factor on the human body, we suggest the method that allows individual evaluation. The main criterion of proving the fact of the interaction of the body with a xenobiotic has been formulated, namely: detecting structural or functional changes in biological systems of the body in combination with identification and detection of chemical markers of such an interaction in biological media.

The successive introduction and approval of the procedure of evaluating of an unknown chemical substance resulted logically in the need for formulating the concept of "xenostatus". "Xenostatus" is a notion that characterizes the state of the interaction of a foreign chemical compound and the body involving entry, presence, accumulation, transformation, and elimination of a xenobiotic. It is obvious that solving the problem of evaluating xenostatus is impossible without solving the problem of evaluating the chemical status consistent with the standard of "chemobiotic status". Solving the problem of the comprehensive evaluation of the human "xenostatus" is complex enough when organic pollutants or their probable metabolites are concerned. As a rule, organic compounds are eliminated out of the body and metabolize efficiently. This being so, interacting compounds proper or their adducts are detected in biological media shortly after the interaction.

The resolution of the problem of identification of typical, as well as persistent, chemical markers of the interaction with chemical compounds in biological media is a top-priority task of theoretical and applied toxicology.

Thus we can say that the suggested procedure of toxico-sanitary evaluation of a complex unknown chemical substance makes it possible to prevent probable negative consequences of the use of extremely hazardous chemical technologies, to provide the appropriate state of health of persons engaged in observing requirements of international agreements of Russia in chemical weapons destruction, as well as populations residing in the vicinity of chemical weapons storage and destruction facilities.

Table 1: Chemical Markers in Biological Media for Determination of Occurrence of Human Exposure to CWA

MARKER	**CWA TYPE**
METHYPHOSPHONIC ACID (MPA)	GB, GD, VX, RVX
O-ETHYL ESTER OF MPA	VX
2-(DIISOPROPYLAMINO)ETHANTHIOL	
O-ISOBUTYL ESTER OF MPA	RVX
2-(DIETHYLAMINO)ETHANTHIOL	
O-ISOPROPYL ESTER OF MPA	GB
O-PINACOLYL ESTER OF MPA	GD
THIODIGLYCOL	
THIODIGLYCOL SULFOXIDE	SULFUR MUSTARD
2-CHLOROVINYLARSONOUS ACID	LEWISITE

CHEMICAL AND BIOLOGICAL-ECOLOGICAL ASPECTS OF RISK ASSESSMENT FOR LEWISITE DESTRUCTION

Leonid Ionov[1], Nikolay Zubtsovsky[1], Lyudmila Makarova[1], Sergey Reshetnikov[1]
[1]*Udmurtia State University, Izhevsk (Russia)*

Abstract: This work gives details on the composition of technical Lewisite and contains recommendations on decontamination of local areas for the case of their contamination with products of Lewisite transformation. There is considered the role of biological monitoring and biological-monitoring investigations in the system of complex environmental monitoring.

Key words: chemical weapon destruction, lewisite, environmental hazard, biological monitoring

In compliance with international agreements Russia has to destroy considerable amounts of chemical warfare agents, including 6.5 thousand tons of Lewisite (β - chlorovinylarsine dichloride).

The degree of risk from the potential release of lewisite into the environment in the process of preparatory operations on dismantling of the containers with chemical warfare agents, filling transportation containers and their transportation to the detoxification facility will increase considerably.

Technical lewisite is a mixture of several arsenious, including:

$ClCH=CHAsCl_2$	$(ClCH=CH)_2AsCl$	$(ClCH=CH)_3As$	$AsCl_3$
α- lewisite (65%)	β - lewisite (to 10%)	γ - lewisite (4-12%)	arsenic chloride(16-21%)
I	II	III	IV

The most toxic are (I) and (IV). They also determine the degree of environmental hazard in the case of their release into the environment - soils, water reservoirs, and atmosphere.

Under natural conditions the chemistry of transformations of (I) and (IV) is due to the hydrolysis and oxidation reactions. The direction and rate of these reactions depend on many factors [1].

V.M. Kolodkin and W. Ruck (eds.),
Ecological Risks Associated with the Destruction of Chemical Weapons, 217–221.

Lewisite hydrolyzes stepwise:

$$ClCH=CHAsCl_2 + 2HOH \underset{fast}{\longleftrightarrow} ClCH=CHAs(OH)_2 + 2HCl \quad (1)$$

V

$$ClCH=CHAs(OH)_2 \underset{slow}{\longleftrightarrow} ClCH=CHAsO + H_2O \quad (2)$$

VI

$$n\,ClCH=CHAsO \underset{slow}{\longleftrightarrow} (ClCH=CHAsO)_n \quad (3)$$

VII

When arsenic chloride hydrolyzes completely, it transforms to arsenious acid (VIII):

$$AsCl_3 + 3HOH \underset{fast}{\longleftrightarrow} As(OH)_3 + HCl$$

VIII

Based on long-term systematic investigations of the chemical behavior of α-lewisite and arsenic chloride (IV) in the samples of various soils and water collected in the lewisite stockpile area and supposed sites of situation of the future facility for lewisite destruction [2] it was found that:

1. The upper soil layers - sandy and podzolic - have weak sorption properties relative to hydrolysis products of α-lewisite and arsenic chloride;
2. The proximity of groundwater to the soil surface occurring everywhere contributes to the fast penetration of hydrolysis products (I) and (IV) into the water-bearing strata and migration of these ecological toxicants to the main water-way, the Kama river;
3. The sorption and salt formation processes which involve β-chlorovinylarsonic acid are insignificant in the soils of the area:
   a) Dehydration of (V) to β-chlorovinylarsine oxide (VI) in the surface soil layers, to a depth of 60-80 cm, proceeds at a low rate and, as a result, lewiste metabolites soluble in water have enough time to penetrate into the water-bearing strata and migrate into water-reservoirs and, finally, into the Kama river;
   b) The conditions for the biochemical detoxification of α-lewisite, arsenic chloride and their metabolites both in the surface layer and water reservoirs are practically absent. In the case of release of

chemical agents into the environment, decomposition and decrease of the concentrations of the toxic forms of arsenic would not take place.

Considering the above, it is possible to outline concrete recommendations on decontamination of local soil areas in the case of their contamination with the products of lewisite transformation [3] and to show the biological-ecological aspects of risk assessment for destruction of lewisite and other arsenious compounds combined with it.

Evidently, biological monitoring (BM) and biological-monitoring investigations (BMI) should occupy a specific place in the system of complex environmental monitoring as a system for collection, processing and storage of information for further administrative decision-making on the environmental protection, prevention of accidents and elimination of their consequences. Note that BM and BMI should be carried out in parallel. We suggest that BMI should be done in three stages:

The first stage: laboratory experiments on determination of the influence of threshold excessive concentrations of metabolites of lewisite on model species of hydrobiontes and ground plants.

The goal of the stage is revealing the most specific biological indicative signs for further practical use in the system of biological monitoring to determine the state of the environment.

The second stage: experiments under field conditions that are maximally close to the soil-climatic and hydro-biological conditions of the area of dismantling, detoxification and destruction of lewisite.

The goal of the stage is to find qualitative and quantitative characteristics of the effects of lewisite metabolites (2-chlorovinylarsin oxide, 2-chlorovinylarsonic acid and respective arsonates) on the selected model objects of the first stage; classification, systematization and mathematical processing of data; establishing laws of accumulation of biomass of the objects at various levels of environmental pollution.

The third stage: breaking ground for monitoring sites in the area of lewisite dismantling and destruction to carry out investigations on the specially selected plots of soil and water areas which are the most typical by the character of biocenoses including model species of bioindicators selected after the first and second stages of investigation.

The goal of the stage is the final selection of bioindicators and determination of their place in the environmental monitoring.

The proposed biological objects for investigations (as potential bioindicators):

1. Hydrobiontes: aquatic plants of ponds, rivers, creeks (aquatic plants, like duckweed, are biological accumulators of arsenic) and animals (Daphnia, Cyclops, Amphibia (larvae)).
2. Soil biota: the most typical plant species specific to the soil landscapes of the area and plants-bioindicators grown as the main crops on the territory of collective and state farms and also in the plots of land attached to the houses of the residential population. The potencial bioindicators in the group of cultivated plants are Triticum, Solanum, Pisum, Avena, Brassica, Daucus, Cucumis and other species possessing various degrees of tolerance to arsenic compounds.

References

[1] Ionov, L.B., Reshetnikov, S.M., Makarova, L.L., Kornev, V.I. (1994). As-containing Chemical Agents. Lewisite and Chemistry of its Transformations. Izhevsk, Herald of Udmurtia University, pp. 4-22.

[2] Ionov, L.B., Reshetnikov, S.M., Makarova, L.L., Zubtsovsky, N.E., Zakharov, V.Yu. (1996). Results of Work on Complex Investigation of Environment in CA Stockpiling Areas of Udmurtia Republic. Second Public Hearing on Problem of Chemical Weapon Destruction. Izhevsk, Herald of Udmurtia University, pp. 51-58.

[3] Kholstov, V.I., Zhdanov, V.I., Kogielev, V.M. (1994). Approaches to Evaluation of Technologies for Chemical Weapon Destruction. Mendeleyev Russian Chemical Journal of Chemical Society, XXXVIII, 2: 42-47.

RISK MAPPING AND RISK ASSESSMENT FOR SUSPECTED CHEMICAL WEAPONS BURIAL SITES ON THE FORMER MILITARY TRAINING AREA DÖBERITZER HEIDE (GERMANY)

Kay Winkelmann[1], Wolfgang Spyra[1], Christine Garbotz[1]
[1]*Chair of Chemical Engineering and Hazardous Wastes, Brandenburg University of Technology, Cottbus (Germany)*

Abstract: The Döberitzer Heide is a former military training area in Brandenburg (Germany) that was used for military training and exercises, military research, testing and development for almost 250 years between 1750 and 1994. The former military training area Döberitzer Heide, situated approximately five kilometres west of Berlin, and other facilities in Germany were used during the development and testing of chemical warfare agents and chemical warfare ammunitions during World War I. Remains of these experiments and tests of warfare agents and ammunitions are still found here today as are burials of disposed chemical warfare agents and chemical warfare ammunitions. On behalf of the Ammunition Clearance Service of the Federal State of Brandenburg, the Havelland District Authority and the site owner, an investigation for suspicious sites for burials of chemical warfare agents and ammunitions is carried out. The approach includes comprehensive historical investigations in national and international archives, the interpretation of historic maps and aerial photographs and remote sensing methods. A geographic information system (GIS) is used for the interpretation of the acquired information. The investigation approach is described in this article.

Key words: chemical warfare, warfare agents, ammunition, UXO, burials

1. INTRODUCTION

A contemporary quote of Prof. Fritz Haber, privy councillor and leading German scientist in the field of chemical warfare agents development and production during World War I on the discovery of large burials containing leaking mustard gas shells in Berlin-Adlershof and the necessary actions gives an impression of the contemporary handling of the destruction of chemical warfare agents and ammunitions:

"In my opinion, the buried shells should best remain untouched for the next 30 years. During this time, the warfare agent in the shells will be destroyed almost completely. Thus, it will be possible to carry out the

V.M. Kolodkin and W. Ruck (eds.),
Ecological Risks Associated with the Destruction of Chemical Weapons, 223–229.

necessary work for the complete destruction of the shells without endangerment of human live.” [1]

Today we know that this approach does not work and does not meet the minimum requirements for human and environmental safety – neither for mustard gas nor for any other chemical warfare agent. And we know that those that were in charge of the development and production of chemical warfare agents are not necessarily those that should be in charge of the destruction of chemical warfare agents. History has shown – not only in Germany – that the history of the destruction of chemical warfare agents is also a history of accidents with losses of human life, health threats for workers and nearby population and environmental damage.

After World War I, the German Army was demilitarised in a mostly disorganized manner. Soldiers abandoned their units and military formations abandoned their equipment, weapons and ammunition. Records show that the German ministry of war and institutions were not aware of the quantity and whereabouts of conventional and chemical warfare ammunition after the ceasefire on November 11, 1918.

Only after the Treaty of Versailles of June 28, 1919 became effective on January 10, 1920, did the German administration start to consider the necessity and the means of the destruction of chemical warfare agents and chemical ammunition. Protocols of the German Reich trust company show that intensive discussion only began in October 1920 [2-1].

Until then, most of the chemical warfare ammunition was stored under unsuitable conditions in Breloh and many other locations in Germany, despite an order of the High Command of the German army of 1918 to transfer all chemical warfare agents and ammunitions to Breloh. While chemical warfare ammunition was to be sent to Breloh, conventional ammunition was to be collected at the nearest artillery depot. Artillery depots were central facilities that were run by the army and that were responsible for the laboration and delaboration of ammunition [2-2, 3].

On October 24, 1919, a huge explosion struck the facilities at Breloh. 48 buildings of the green and yellow cross production facilities were destroyed, about one million shells filled with chemical warfare agents, one million detonators, 230,000 mines, and 40 tank wagons for chemical warfare agents were destroyed. Parts of shells, tanks and buildings were scattered over an area within a radius of three kilometres around the facility [4].

While the clean-up and destruction at the Breloh facilities began in a mostly organized manner (while not environmentally compatible or according to occupational health standards) following the disaster on October 24, 1919, reports show that it was still unclear what quantities of chemical warfare agents and chemical warfare ammunition remained in other places in Germany, mostly at artillery depots that were in charge of the delaboration of conventional ammunitions. And the reports also show that, where such stocks where known, they were not always handled properly, not safe for transport, or no safe (railway) means of transport were available. And, due to improper storage of chemical warfare ammunition, the coloured markings (crosses) were not always recognizable on the ammunitions. Thus, in some cases, it was only recognized during the delaboration that a certain batch of ammunition was chemical warfare ammunition that was not recognized before and thus not brought to Breloh [2-3, 5]. Several reports show that in some of these cases, chemical warfare ammunition was buried in the vicinity of the artillery depots or delaboration facilities [6, 7]. But also during the production of chemical warfare agents and the laboration of chemical warfare ammunitions, the burial of leaking shells or excess production occurred because no other means of destruction were known or available [8].

2. PROBLEM

The former military training area Döberitzer Heide is situated five kilometres west of Berlin and was used as a military training area for about 250 years between 1750 and 1994. Because of its vicinity to Berlin, it was not only used for extensive military training, but also for development of testing of new weapons. One of the scientific institutions that used the Döberitzer Heide for testing purposes was the Kaiser-Wilhelm-Institute for Physical and Electrochemistry in Berlin-Dahlem. Between 1915 and 1918, the institute under its director Prof. Fritz Haber was the leading scientific institute in the development of chemical warfare agents and chemical warfare ammunitions in Germany. The Döberitz military training area was used for testing of shells containing chemical warfare agents. The shells were typically detonated in free areas, trenches or dugouts and the toxicity and effects tested in animal experiments and/or by human volunteers [9, 10].

After 1945, the 50-square-kilometer military training area was used by Soviet troops until 1994. In 1994, the largest part of the Döberitzer Heide became a nature preserve while five square kilometres remain a military training area to be used by the German Bundeswehr. During ammunition clearing operations in preparation of the use by the Bundeswehr, several burials containing chemical warfare ammunitions were found, some in systematic burials, others in groups of three or four in unsystematic burials.

3. INVESTIGATION APPROACH

In order to assess the probability of further burials containing chemical warfare agents and ammunitions, and to detect possible burials, an investigation on behalf of the Ammunition Clearance Service of the federal state of Brandenburg, the Havelland District Authority and the site owner is carried out.

The investigation approach includes a historic investigation in all relevant and accessible national and international archives, the multi-temporal interpretation of historical aerial photographs and maps for the timeframe 1900 to 1994, and remote sensing methods, in particular high-resolution multispectral digital aerial photographs and night-time thermal remote sensing. The objective is to derive a risk map for the former military training area Döberitzer Heide with respect to burial sites and test sites for chemical warfare agents. Due to the investigation approach, the map will also includes hazards that are due to conventional ammunition (shooting ranges, destruction and blasting areas) and military use (burials of wastes, etc.).

3.1 Historic Investigation

The historic investigation targets the practice and state of the art of contemporary handling, use and destruction of chemical warfare agents in general and – as far as possible – the actions carried out on the former military training area Döberitzer Heide with respect to chemical warfare agents, military training and exercises. The historic investigation focuses on records of fiscal and military authorities, contemporary scientific literature on chemical warfare and warfare agents, staff known to have been working in the field of chemical warfare agent development, rules for military training areas, and military service regulations. A second part of the historic investigation focuses on historic maps and aerial photographs.

The objective of the historic investigation of records and literature is to assess if, when and in what quantities chemical warfare agents were handled in the area of the former military training area Döberitzer Heide. The objective of the retrieval and interpretation of historic maps and aerial photographs is to determine when and where chemical warfare agents might have been handled. Information from archival records is used for the interpretation of maps and aerial photographs, because no direct information on the time and place of the use of chemical warfare agents is recorded neither on maps, nor aerial photographs. Restricted areas, areas where entrenchment works were forbidden or a stand of trees of a certain age can give indications together with the (oral) information that around 1915-17 a military service regulation existed that prescribed to bury chemical warfare agent shells at a depth of about one metre, lay out the pit with chlorinated lime, put in the shells in order, cover them with chlorinated lime and earth and plant oak trees on the burial. The comprehensive and contextual interpretation of maps and aerial photographs in their spatial relations to other features of interest such as scientific, production/laboration, and delaboration facilities also yields areas, that were excluded from military training and exercises for long times without obvious reason. Historical aerial photographs may also show areas that were excluded from use for certain times or areas of disturbed earth that may direct to burials. However, aerial photographs are only available for the area from 1941 and only in time phase of five to ten years, thus not every relevant object/area can be reveale Those historical aerial photographs that are available in stereo pairs will also undergo visual stereoscopic analysis for terrain anomalies that might be connected to burials.

3.2 Remote Sensing

The remote sensing part of the investigation includes high spatial resolution multispectral digital aerial photographs using a ZI/Imaging DMC camera system and nighttime thermal imagery using a FLIR Systems ThermaCAM PM695. The multispectral digital aerial photographs were acquired in June 2003 and have been processed to become digital orthophotos. They will be investigated for anomalies in the vegetation structure, terrain anomalies and vegetation stress. Furthermore, the aerial photographs are being used for the geo-referencing of historic aerial photographs and maps. They are also the basis for mapping of the area of interest. The thermal imaging operation that was carried out in August 2003 aims directly at burials. In previous laboratory and field investigations, it has been shown that thermal images can show burials by revealing temperature differences between disturbed and undisturbed soil. Because of increased water capacity and soil moisture, disturbed soil typically shows radiation temperatures that are 0.5 °C to 1.5°C lower than those of undisturbed, similar neighbouring soils. Field trials during the thermal campaign in August 2003 conducted as a test have also shown that increased soil moisture in the subsurface is clearly visible in the thermal imagery. A previous thermal campaign in 1998, applying military equipment under unfavourable external conditions, was not successful [11,12].

Because of the sensitivity for areas of increased soil moisture and other subsurface inhomogeneities, the thermal imagery interpretation will yield information on burials in open (not forested) areas, thus being insensitive to the contents or age of a burial.

4. CONCLUSIONS

All information retrieved from historic and remote-sensing investigation means, together with field studies, is being geo-referenced and integrated in a geographic information system (GIS). The system will integrate all spatial information sources that can be viewed and analysed in the system. An algorithm that is currently in development will be used for a hazard analysis for certain spots or areas in the area of investigation with respect to chemical warfare agents/ammunitions, risk posed by conventional ammunition and buried or deposited wastes. A risk map with a sound basis for the former military training area and its facilities will be the result of the investigation. The investigation will provide a risk assessment and a risk map mainly based on non-contact investigation means.

References

[1] Quotation of Fritz Haber, Memorandum dated 17. February 1923 „Nach meiner Meinung bleiben die vergrabenen Geschosse am besten während der nächsten 30 Jahre unberührt liegen. In dieser Zeit würde der Gaskampfstoff in den Geschossen zumeist zerstört sein, sodass die zur völligen Vernichtung der Geschosse dann noch erforderlichen Arbeiten dann ohne Gefährdung von Menschenleben ausgeführt werden könnten." German Federal Archives Berlin, R2201/3305.

[2] Minutes of a meeting on the destruction of chemical warfare agents dated Nov. 4, 1920 held in the German Treasury Department (Reichsschatzministerium). German Federal Archives Berlin, R2201/3305.

[3] Demobilmachungsplan für das Deutsche Heer, Berlin 1918. Demobilisation Guidelines. German Federal Archives (Military Archives) Freiburg. AIII8.

[4] Hamerla F. Chronik der Raubkammer. Munster, October 1973.

[5] Hartley H. Report on German Chemical Warfare Organisation & Policy, 1914 – 1918. Public Record Office, London. WO 33/1072.

[6] Contract for the disposal of chemical ammunitions. German Federal Archives Koblenz. R133/116.

[7] Letter of First Lieutenant (ret.) Bähnsch to the Fiscal Authorities Wittenberg, dated Dec. 7, 1922. German Federal Archives Berlin, R2201/3305.

[8] Memorandum of a visit of the burial sites of leaking mustard gas shells in Berlin-Adlershof by Mr. Ehrt, dated Nov. 4, 1920. German Federal Archives Berlin, R2201/3305.

[9] Documentation of animal experiment in Döberitz, dated May 5, 1917. Germany Federal Archives (Military Archives) Freiburg, RH12-9-26.

[10] Große Kampfstoffmappe des KWI. Germany Federal Archives (Military Archives) Freiburg, RH8-1870 – RH8-1876.

[11] Winkelmann K, Spyra W. 2001. Thermalaufnahmen zur Erkundung ehemaliger Truppenübungsplätze. Photogrammetrie – Fernerkundung – Geoinformation. 6/2001. 417-426.

[12] Irvine JM et al. 1997. The detection and mapping of buried wastes. International Journal of Remote Sensing. 18 (7). 1583-1595.

ABOUT EVALUATION PRACTICE OF EMERGENCY ENVIRONMENTAL RISKS OF CHEMICALLY DANGEROUS FACILITIES

Gennadiy Arbusov[1], Boris Laskin[1]
[1]*Federal State Unitary Establishment Russian Scientific Center "Applied Chemistry", Saint-Petersburg (Russia)*

Abstract: Evaluation of operational risk of dangerous industrial facilities in Russia is made on the basis of standard and scientific studies in the field of environmental and industrial safety. Using this basis in the report, some problems are examined, which are stressed by the ecological appraisal of planned or operational chemically dangerous facilities (abbreviated CDF), and also by practical declaration of industrial CDF safety.

Key words: ecology, accidents, risks, evaluation

1. INTRODUCTION

At a sufficiently thorough study of the basic theses of evaluation of CDF operational risk there exist a number of aspects of such evaluation, in which an uncertainty in the evaluation of some indicators remains.

With reference to the environmental risks the primary of such aspects are:

- lack of conventional environmental risk determination and of standard approaches in the environmental risk evaluation,
- lack of common approaches in the economic evaluation of human losses and sanitary lesions both at the direct injurious damage effect and at the subsequent effect through recipients in polluted nature ,
- lack of engineering methods in the rules of summation of risks in different nature, including ecological risks for different recipients of environment and biota,
- lack of common approaches in determination of acceptable risk levels.

The presence of such problems, and uncertainties produced by them, prevents precise evaluation of total risks of CDF operation.

Taking into account the current development condition of scientific and technologic base of Russia, we shall consider the possible analysis directions of these problems.

V.M. Kolodkin and W. Ruck (eds.).
Ecological Risks Associated with the Destruction of Chemical Weapons, 231–240.

2. RISK EVALUATION PROBLEMS ANALYSIS

2.1 Analysis of approaches to the environmental safety and environmental risks

In the practice of environmental safety level of the region where CDF's are present on its territory evaluation methods are usually applied, of ecological and economic damage (e.g. [1, 2]) and environmental risk (e.g. [3, 4, 5]). The latter does not imply the risk of deterioration of nature quality, but the risk of population and personnel health impairment resulting from contacts with affected recipients.

The development of the application of methods of environmental and economic evaluation of damage from CDF is concerned with the fact that environmental damage usually cannot be adequately measured using material damage categories or other indications.

These conceptions are mostly developed in the works of Russian scientists I.B. Evstafiev, V.I. Kholstov, et al. (e.g. [6]), V.G. Gorsky and colleagues (e.g. [7]).

Environmental risk evaluation in Russia remains not only partially-developed. According to [8], it is envisaged, that the the development of the risk evaluation use programmes in Russia will be complete only by 2005. Strategy recommendations for Sverdlovsk region "Evaluation strategy of environment pollution risk for the health of population" [18] up to now have not been disseminated in other subjects of the federation. Therefore for environmental evaluation of new construction it is better to apply the methods of potential evaluation of risks, based on local principles of hygienic normalization [10].

As indicators of standards, which are applied in the evaluation of emergency environmental risk, values of payments are considered for pollution, accepted in the legislative and statutory acts [11, 12] in the area of atmosphere pollution, [11] – in the area of marine recipients pollution, [11, 12, 13] – in the area of soil pollution, [14, 15] – in the area of forest protection, [16, 17] – in the area of animal world protection, [18, 19] – in the area of fish resources protection.

The evaluation of averted ecological damage to the environment is carried out in accordance with the methods approved by the State Environmental Protection Committee of Russian Federation [1].

2.2 Analysis of the problem of application of the economic evaluation methods of human losses and medical injuries

Risk evaluation methods recommended today by the injuries of personnel and population as a result of accidents are based on calculation of individual and collective risks.

Underlying the basis of economic evaluation of human losses, which occur both as direct injuries of people (as results of explosions, conflagrations, poisonings) and delayed injuries after contact with affected recipients in nature, is lies an approach which consists of multiplying the number of untimely deaths by the average statistical economic value of diminished life.

Such evaluations in economically developed countries show [3] that the value of lost years of remaining life (accepted equal to 30 as the average number of years of remaining life for all population groups) is evaluated in the range from 20000 to 450000 US dollars (average statistical value is 160000 US dollars).

Considering the substantially lower average per capita revenue of Russian population, it is recommended to accept the lowest value of average statistical emergency risk value for one additional death in Russia, equal to 3000 US dollars.

The average price of risk of death in an accident may exceed the amount of pecuniary compensations customary in Russia in the presence of insurance responsibility before personnel and third persons, personal insurance contracts, or legal action payments.

The compensation level of medical injuries as consequences of accidents are determined by the active legislation in the field of protection of labour but average statistical values of economic equivalent of the damage resulting from injuries are recommended to be adopted at 10% of the amount of average statistical values of economic equivalent of human life.

2.3 Analysis of summation of risks concerned with different damage types

In conformity with accidents which result in homogenous losses (i.e., affects only personnel and population or only material damages, etc) the calculation methods of such risks are well developed and are commonly applied in practical evaluation of CDF risks. However evaluation methods of complex risk types are not yet perfected or do not exist. To these types of risk we can attribute direct and indirect environmental risks.

Considering the accepted environmental and economic concept of evaluation of emergency environmental risks for specific CDF's, and accounting for individual economic constituents of damages in the common value of these risks ,which damage results to personnel and population of residential areas from affected recipients in nature in postdamage periods, summation of environmental risks is made in the following way:

- according to the measure of changes in the environmental situation for the surrounding environmental recipients, which is evaluated in accordance with the standard methods [20];
- according to the damage caused to environmental recipients and biota, in the postdamage period in accordance with the methods [1];
- according to the damage concerned with the pollution of environmental recipients and caused to personnel and population in the postdamage period, determined in accordance with the above stated economic indicators.

In this case, the total value of environmental risk for the period in question is summed up according to certain rules from all risk types evaluated for all scenarios of development of events of all environmental damages.

2.4 Analysis of the problem of evaluation of risk acceptability

Russian developments in this area are about 10 years old and their publication dates from the period of the beginning of chemical weapons destruction works in Russia (e.g. [6, 21, 22]).

The basic concept of domestic methods of evaluation of acceptable risk level, as well as in the developed countries, is based on the evaluation of acceptability of individual risk level.

For every geographical region of Russia usually some health factors are available that influence the state of human health in one or another way and accordingly they influence the "death which resulted from unnatural reasons". These circumstances determine the existence in this region of "background risk" which is related to the category of "antropogenic risk". The scale of acceptable risk in the evaluation territory may differ from local (urban) to territorial and regional. The background risk indicators in Russia greatly exceed those in developed countries.

3. PRACTICE OF COMPLEX EVALUATION OF EMERGENCY RISKS

The practice of evaluation of environmental and economic safety level of analyzable planned or operational CDF is formed with accounting for the territory zonation rules according to criteria of risk for the population, and also on the work of the authors of this report together with V.G. Gorsky and G.A. Motkin in accordance with the CDF ranking for point and line sources of danger according to their technological indicators. (e.g. [23]).

The evaluation of the extent of existing danger (risk) during CDF operation is conducted with the help of the recommendations made by the working group No. 12 of UNESCO and State Urban Technical Supervision of Russia matrix of risk level ranking (Table 1).

In Table 1 risk levels are marked A, B, C, D (high, medium, low and standard risk levels correspondingly). Risk level categories are determined by the frequencies of accident occurrence and scope of damage resulting from them. The acceptable is considered the risk level with the category "standard". The categorization of accidents for such evaluations is conducted with consideration of emergency situations classification, which is worked out by the Ministry of Emergency Situations of Russia [26].

Table 1. Matrix of risk level ranking [24, 25]

Frequency of occurrence	Indicative frequency per year $^{-1}$	**Severity of consequence**			
		Catastrophic	**Major**	**Severe**	**Minor**
Frequent	>1	A	A	A	C
Probable	$1\text{-}10^{-2}$	A	A	B	C
Remote	$10^{-2}\text{-}10^{-4}$	A	B	B	C
Improbable	$10^{-4}\text{-}10^{-6}$	A	B	C	D
Incredible	$<10^{-6}$	B	C	C	D

Taking into consideration the above-stated positions RSC "Applied chemistry" has carried out the Risk Analysis of point and line sources of danger of one of the operating CDF's in North-West region of Russia which utilizes liquid chlorine in its activity.

In the course of the analysis are determined the consequence severity levels at the possible accidents in this particular CDF (Table 2).

Table 2. Classification of scale of environmental effects on CDF

Scale of effects	Characteristics of effects
Catastrophic	Release of toxic products located on the territory of CDF, during which the depth of lethal injury zone spreads beyond the territory of the damage area, and the zone of medical injuries exceeds the responsibility zone of the CDF and may reach residential zone. The damage is not less than $7x10^6$ Euro.
Major	Release of toxic products, during which the lethal injury zones are located within the bounds of the damage production sector, and the zones of medical injuries are located within the bounds of the sanitary zone of the establishment. The damage is in the range $6x10^5$-$7x10^6$ Euro.
Severe	Release of toxic products, during which the medical injury zones cover the area of damage production sector. The damage is in the range $1.3x10^4$- $6x10^5$ Euro.
Minor	Release of toxic products, during which the medical injury zones are located within the bounds the workplace of maintenance staff in the damage area. The damage is less than $1.3x10^4$ Euro.

The risk levels are classified for the same CDF (Table 3).

On the assumption of the accepted original positions the main indicators of risk are determined for the scenarios of damages that bring maximum danger (Table 4).

The evaluation of ecological and economic damage to natural environment recipients was conducted separately for each environment.

Table 3. Characteristics of environmental risk levels for CDF.

Risk level	Characteristic of risk	The conditions of occurrence of risk
A	High	As a rule, with damage frequency exceeding $1.0x10^{-6}$ times a year, with disastrous effects, and also with damage frequency not below than $1.0*10^{-2}$ times a year for damages with critical effects.
B	Intermediate	As a rule, with damages that have frequencies $1.0*10^{-2} \div 1.0*10^{-6}$ times a year, with critical effects, and with frequencies $1\text{-}1.0*10^{-4}$ times a year for damages with uncritical effects.
C	Low	As a rule, with damages that have frequencies exceeding $1.0*10^{-6}$ times a year, with uncritical effects, and with frequencies exceeding $1.0*10^{-4}$ times a year for damages with negligible effects.
D	Trivial	As a rule, with damages that have frequencies below $1.0*10^{-4}$ times a year, with negligible effects.

Table 4. The results of the analysis of the risk of CDF activity (values of 01.01.2003)

Types of damage	Frequency of the most serious damage, times a year	Total damage, thousands of Euro	Average annual damage, thousands of Euro	Risk level
Environmental	$1.7x10^{-3}$	61	0.104	Low
Collection	$1.6x10^{-5}$	528	0.0085	Low
Material	$1.7x10^{-3}$	0.371	0.006	Low
Total value		589.4	0.1185	Low

For determination of value of averted damage from atmospheric emission of contaminants, average design values of damage for a unit of reduced mass of atmospheric contaminations are used.

For the damage evaluations, types of violations stated in the Federal law "About subsoil" were taken into consideration.

For the evaluation of damage resulting from contamination of marine recipients, data about payments for disposal of sewage water according to [13] was used.

Included in the damages to the plant world are considered the expenses for fire preventive measures organization (fire barriers) in forests, adjoining to building or operation sites of CDF, as well as the expenses for reforestation.

The basic predictable damage to the animal world is postulated from the decrease of nesting (reproduction) density of birds and other surface vertebrates as a result of animal deterrence or direct extermination.

The damage to the fish resources mostly consists of two types: spawning ground losses and loss and production decrease of forage resources. The amount of expenses, which are necessary for compensation of predicated damage to fish resources, is included into the fish resources damages content.

On the assumption of the produced results follows that the risk of activity of the CDF in question is acceptable considering the current organization of manufacturing process (Table 4).

4. CONCLUSION

Thus an engineering CDF risk level evaluation is proposed on the basis of research works of Russian and foreign scientists and normative materials.

References

[1] Methods of evaluation of averted ecological damage, approved by the State Committee of Environmental Protection, 30.11.99, M., (1999), 71p.
[2] Strategic instructions on environmental risks evaluation, (2000) N.Novgorod.
[3] Bykov, A.A, Solenova, L.G., Zemlyanaya, G.M., Furman, V.D. (1999). "Strategic recommendations on analysis and management of risk of environmental hazards effect on the health of population", M.
[4] Onischenko, G.G., Novikov, S.M., Rakhmanin, Y.A. (2002). Basics of evaluation of risk to the health of population at the impact of chemical agents, which contaminate the environment, M., RI of ES and GOS.
[5] Kiselev, A.V., Fridman, K.B. (1997). Evaluation of health risk, S-Pb, 104 p.
[6] Evstafiev, I.B., Kholstov, V.I., Grigoriev, S.G. (1993). "Strategic basics of emergency situation evaluation of the facilities of storage and determination of chemical weapons", Chemistry Journal (named after D.I. Mendeleyev), XXXVII, N3.
[7] Gorsky, V.G., Shvetsova-Shidlovskaya, T.N., et al (2003). Methodology of risk forecast of chemical manufactures in conditions of source information uncertainty, collected reports of All-Russian conference, v.2, S-Pb, 163-168.
[8] "About use of risk evaluation methodology for environment quality management and health of population of Russian Federation management". Decree of the Administration of Sverdlovsk Region, approved by the Chief State Sanitary Doctor of the Russian Federation on 10.11.97, No. 25, the Chief State Inspector of the Russian Federation of Nature protection on 10.11.97, No. 03-19/24-3488.
[9] Methodology of evaluation of environment contamination risk for health of the population. Risk evaluation and ecologic epidemiology". Ekaterinburg, 2000, p. 34-75.
[10] Evaluation of potential risk to health in the system of hygienic monitoring at the evaluation of environment quality. http//:home.comset.net/anatkiss.methrisk.htm
[11] Decree of the Russian Federation Government dated 28.02.92 No. 632 About approval of payment determination and its limit amount for contamination of natural environment, disposal of wastes, other types of deleterious effect.
[12] Basic standards of payments for discharge, disposal of contaminants into the environment and disposal of waste products. Approved by the Ministry of Nature of Russia on 27.11.92.

[13] "About approval of minimum and maximum payment rates for use of marine recipients in the river, lake and sea basins". Decree of the Russian Federation Government dated 22.07.1998, No. 818.
[14] Russian Federation Act "About land charge" dated 11.10.91, No. 1738-1 ФЗ.
[15] Order of evaluation of damage due to pollution of soil by contaminants. Approved by the Ministry of Nature of Russia on 18.11.93.
[16] Forest Code of the Russian Federation of 29.01.97, No. 22- ФЗ.
[17] "About ratification of fixed rates for calculation of levy amount for the damage caused by illegal obtainment or annihilation of animal of plant world recipients", order of the Ministry of Nature of Russia of 04.05.94, No. 126.
[18] Russian Federation Act "About animal world" of 24.04.95, No. 52- ФЗ.
[19] "About ratification of fixed rates for calculation of levy amount for the damage caused by annihilation, illegal catch or obtainment of water biological resources recipients", Decree of Russian Federation Government dated 25.05.94, No. 515.
[20] Methods of toxic dozes calculation and probabilistic forecasting of affection by violent poisonous substances, M., RI SOES, 1993.
[21] Elokhin, A.N. (1999) "Recommendations of acceptable risk determination, in collected articles "Methodological aspects of man-caused and natural risks evaluation", M., 64-75.
[22] Kharisov, G.K (1993). "Substantiation of inputs provided for averting of people death at accidents, damages, catastrophes, acts of God", journal "Safety problems at emergency situations", issue 8, p. 73.
[23] Arbuzov, G.M., Laskin, B.M., Gorsky, V.G., Motkin, G.A. (2000). Evaluation of chemical danger of an enterprise in the system of insurance ecological audit. Theses of the conference "Theory and practice of ecologic insurance" Kaliningrad, 15-28.
[24] Guidelines for the risk analysis of technological systems. WG 12, H:/OSD9B/J37/03.94.
[25] Methodical guidelines for realization of analysis of dangerous industrial facilities risk. РД 03-418-01. Approved by the Decree of Municipal Engineering Supervision of Russian of 10.07.01 N30.
[26] About classification of emergency situations of natural and man-caused nature. Decree Russian Federation Government dated 13.09.96 N 1094.

RISK ASSESSMENT AND SAFETY MANAGEMENT AT CHEMICAL FACILITIES WITH THE USE OF NEW INFORMATION TECHNOLOGIES

Alexander Egorov[1], Tatiana Savitskaya[1]
[1] *D.I. Mendeleyev Chemical-Technological University, Moscow (Russia)*

Abstract: This work proposes to use a qualitatively new type of automated system on the basis of up-to-date information technologies to enhance the safety of chemical facilities: integrated computer-aided control systems (ICACS) combining into one structure information-modeling and control systems, program complexes. To ensure the environmental safety of chemical facilities, the creation of ICACS to test and control the quality of the free air over the territories in the immediate proximity to chemical facilities is proposed. To enhance the industrial safety of chemical facilities, an integrated computer-aided system of risk assessment and safety control has been developped. To exercise the analysis and risk assessment of various chemically dangerous facilities, logical-graphic and probabilistic models of risk assessment are proposed.

Key words: risk assessment, safety management at chemical facilities, integrated computer-aided control systems, environmental safety, industrial safety, decision-making support system

The problem of safety at chemical facilities was studied by many noted national and foreign scientists, among whom the academicians Legasov V.A. [1] and Kafarov V.V. [2], professors Gorsky V.G. [3, 4], Marshall V.[5], Henly E [6], Kletz T. [7] deserve mentioning.

Chemical facilities are among the most dangerous technogenic sources of impact. This is due, first of all, to regulated discharges of pollutants into the free air and the discharge of a considerable amount of un-recycled waste products into water reservoirs and the soil.

The danger of chemical facilities doubles up in the case of emergencies associated with their operation and also during storage and transportation of dangerous chemical agents (DCA).

It is very important to ensure a minimum level of risk of accidents at chemically dangerous facilities and risks of direct or indirect effects of dangerous chemical substances on the environment and humans in the process of creation of safe chemical facilities.

This work proposes to use a new type of automated system on the basis of state-of-the-art information technologies to enhance the safety of

V.M. Kolodkin and W. Ruck (eds.),
Ecological Risks Associated with the Destruction of Chemical Weapons, 241–246.

chemical facilities: integrated computer-aided control systems (ICACS) combining into one structure information-modeling and control system, program complexes and technical means for data collection and transfer on the basis of local computer networks.

To ensure the environmental safety of chemical facilities, we proposed creation of ICACS to test and control the quality of the free air [8], the main functions of this testing being the following:

- analysis of the state of the air over the territories in the immediate proximity to chemical facilities and the sanitary-protective zone;
- prediction of tendencies of its pollution both with continuous sources of discharge of facilities and as a result of accidents with releases of dangerous chemical agents as;
- identification of sources of pollution of the free air and decision-making on the air quality management.

To enhance the industrial safety of chemical facilities, an integrated computer-aided system of risk assessment and safety control (ICACS of safety) [9, 10] is developed. The main purpose of this system is to ensure the industrial safety of the chemical facility, i.e., monitoring industrial dangers, risk analysis and assessment, and making decisions on safety management.

The technical approach and principles of construction of integrated computer-aided control systems have a common methodical basis, which is a systematic analysis.

The major functional components of ICACS are the following: information sub-system, information-modeling sub-system, and decision-making support system (DMSS) for the industrial and environmental safety management.

The authors proposed methodical aspects of creation of integrated computer-aided safety control systems of chemical facilities in [9]. This work presents the results of creation of the functional structure of the ICACS of safety of chemical facilities and its sub-systems and proposes a functional structure of the decision-making support system on the safety management at chemical facilities.

An information sub-system is a distributed database of ICACS comprising a collection of databases on:

- the state of technological processes in the regime of normal operation;
- the possible failures of the equipment, instrumentation, and process technology automated control system, and the breakdown in process and remedies;
- known accidents at facilities and evaluations of their aftereffects;
- real-time databases for collection of information transferred from microprocessor distributed systems of monitoring and control over

specific chemical-technological processes and chemical-technological systems and this information transfer to the sub-system for analysis of production dangers for monitoring the state of the chemical-technological processes, equipment and process technology automated control system.

The main functions realized in the module for analysis of production dangers and risk assessments, in the information-modeling sub-system of the ICACS of safety of chemical facilities are the following:

- identification of potential dangers, revelation of accident situations, risk factors, and analysis of scenarios of development of accident situations;
- quantitative assessment of various types of risk relative to all possible scenarios of development of accident situations;
- ranking accident situations by the degree of their danger.

To obtain a formalized presentation of the analysis and risk assessment technique, the authors proposed to use logico-graphic models in the form of causal-investigatory interrelationships [9, 11]: failures-scenarios-factors-risks and probabilistic models of risk assessment. Such models were created by us for various chemically dangerous facilities and a number of them (models and algorithms of Risk Analysis from periodic chemical productions and large-capacity continuously functioning facilities) were published in [9, 11].

Previous work [9] proposed a model of quantitative risk assessment as a probabilistic relation. The probability of the j-th scenario due to the u-th failure is determined by the following relationship (P_j):

$$P_j = 1 - \prod_{u=1}^{U} (1 - P_{ju}), u = \overline{1, U}, j = \overline{1, M} \quad (1)$$

where P_{ju} is the probability of the u-th failure leading to the j-th scenario; M is the number of scenarios; U is the number of primary failures resulting in the j-th accident situation.

The probabilities that the j th accident situation will lead to one of the risk factors of the same level are determined for each accident situation that develops by the multi-level scenario from the following relationship:

$$F_{ij} = (1 - \sum_{l=1}^{N-1} F_{lj}),\ l \neq i,\ i = \overline{1, N} \tag{2}$$

where F_{ij}, F_{lj} are the probabilities of the *i*-th or *l*-th risk factors of the same level due to the *j*-th accident situation.

The probability that the originated *i*-th risk factor of the *j*-th situation of F_{ij} will lead to the following development of the accident by the *g*-th scenario for any *i'*–th level ($i' = \overline{1, I'}$) of development of the accident ($E^k_{i'_g ij}$) is determined by the relationship:

$$E^k_{i'_g ij} = (1 - \sum_{g=2}^{G} E^k_{i'_g ij}),\ i' = \overline{1, I'},\ i = \overline{1, N},\ j = \overline{1, M} \tag{3}$$

where I' is the number of levels of development of the accident from the *i*-th factor; G is the number of possible scenarios of development of the accident from the *i*-th factor and the *j*-th accident situation. Here the index i'_g characterizes the level number of g-th scenario. Note that various scenarios of development of the accident ($g = \overline{1, G}$) due to the same risk factor (F_{ij}) can have a varied number of levels.

The probability of the *k*-th type of risk from the *j*-th accident situation, and *i*-th risk factor initiated the multi-level development of the accident ($i' = \overline{1, I'}$) by the g-th scenario (R^k_{ijg}) is determined by the following relationship:

$$R^k_{ijg} = P_j \times F_{ij} \times \prod_{i'=1}^{I'} E^k_{i'_g ij},\ k = \overline{1, K},\ g = \overline{1, G}, \tag{4}$$

where *K* is the number of types of risk (environmental, economic, and social).

The probability of the *k*-th type of risk in the *j*-th situation developing in the case of any g-th scenario is determined by the following relationship:

$$R_j^k = P_j \sum_{g=1}^{G} \left(F_{ij} \prod_{i'=1}^{I'} E_{i'_g ij}^k \right), \qquad (5)$$

$i = \overline{1,N}$, $j = \overline{1,M}$, $k = \overline{1,K}$

The probability that risk of the *k*-type will definitely occur during occurrence of at least one scenario is computed by the relationship:

$$R^k = 1 - \prod_{j=1}^{M} \left((1 - P_j \sum_{g=1}^{G} \left(F_{ij} \prod_{i'=1}^{I'} E_{i'_g ig}^k \right) \right) = 1 - \prod_{j=1}^{M} (1 - R_j^k), \qquad (6)$$

$k = \overline{1,K}$, $i = \overline{1,N}$

The proposed model is intended for analysis of the risk of accidents at chemically dangerous facilities developing by various scenarios.

Ranking the accident situations by their contribution to the integral risk index is made with the aim of revealing the most dangerous scenarios and direction of risk reduction by means of technological and organizational measures for the safety management at chemical facilities.

To solve the problems of analysis of the production process dangers, in risk assessment and safety management at chemical facilities, we have developed the ICACS software installed in the client-server architecture [10].

The developed methods, models and program complexes for analysis, risk assessment and safety management were applied to solution of practical problems for large-capacity continuously operating facilities, multi-assortment chemical facilities, and technological equipment with dangerous chemical agents, and working out recommendations on the safety management at chemically dangerous facilities.

References

[1] Legasov, V.A. (1996). From Today to Tomorrow . Thoughts Aloud. Progress, M.

[2] Kafarov, V.V, Meshalkin, V.P., Grun, G., Nojman, V. (1897). Provision and Methods of Optimization of the Safety of Chemical and Petroleum-Refining Facilities. Khimiya, M.)

[3] Gorsky, V.G., Kurochkin, V.K., Djumayev, K.M., Novoseltsev, V.N., Braun, D.L. (1994) Risk Analysis: Methodological Basis of Provision of the Safety of Chemical Technological Facilities. Rossijskij Khimicheskij Zhurnal 2: 54-61.

[4] Gorsky, V.G., Shvetsova-Shilovskaja, T.N., Tereshtshenko, G.F., Putilov, A.V. (2002). Methodology of Analysis of the Accident Risk at Chemically Dangerous Facilities. Khimicheskaia Tekhnologija10: 7-13

[5] Marshall, V. Major Dangers of Chemical Facilities. (1989). Edited by Chajvanov, B.B. and Chernoplekov A.N. Mir, M.

[6] Khenli, E.D., Kumamoto, Kh. (1984) Reliability of Technical Systems and Risk Assessment. Mashinostroenie. M.

[7] Kletz. (1991). An Engineer's View of Human Error. Institution of Chemical Engineers.

[8] Egorov, A.F., Savitskajia, T.V. (2002). System Analysis, Risk Assessment and Safety Management at the Facilities of the Chemical and Related Sectors of the Industry. Khimicheskaya Tekhnologiya 10: 14-22.

[9] Egorov, A.F., Savitskajia, T.V. (2002). System Analysis, Risk Assessment and Safety Management at the Facilities of the Chemical and Related Sectors of the Industry. Khimicheskaya Tekhnologiya 10: 14-22.

[10] Egorov, A.F., Savitskaja, T.V., Filippova G.V. (2002) Up-to-date State in the Field of Risk Analysis, Assessment and Management in Emergencies of the Technogenic Character Associated with Accidents at Industrial Facilities. Environmental Expertise. VINITI Review Information 1.

[11] Egorov, A.F., Savitskaja, T.V., Makarova A.S. (1998). Creation of the Models of Risk Assessment for the Facilities of the Chemical Industry. Khimicheskaya Promyshlennost' 7: 55-63.

ON THE ANALYSIS OF ENVIRONMENTAL RISKS ASSOCIATED WITH THE POSSIBLE LEAKAGE OF CHEMICAL WARFARE AGENTS DURING TRANSPORTATION AND DISPOSAL OF MUNITIONS

Leonid Vasilyev[1]
[1]*Research Institute of Natural and Technogenic Disasters, Udmurtia State University, Izhevsk (Russia)*

Abstract: The work shows that environmental risks associated with various mechanisms of leakage of chemical warfare agents (CWA) into the environment at small temperature variations are probable in the conditions of storage, transportation and destruction of the CWA-filled munitions. A considerable degree of integrity loss in munitions becomes possible during heating to several tens of degrees higher than the storing temperature due to the plastic deformation of the casing. Heating to higher temperatures leads to a complete loss of integrity by means of the destruction of the shell parts. Relevant computation formulas for the quantitative assessment of the consequences of the thermal effect are given in the present contribution.

Key words: environmental risks, leakage of chemical warfare agents, transportation and destruction of chemical weapon

1. INTRODUCTION

Thermal and mechanical factors due to unexpected changes in the external conditions are possible during storage, transportation and destruction of the CWA-filled munitions. As a rule, the vessels with CWA are well protected against direct mechanical actions. The degree of protection against thermal actions is, usually, very much lower, as the transmission mechanisms and the efficiency of the thermal and mechanical actions differ substantially. An insignificant (from the mechanical point of view) thermal increase of gaps in the gaskets and threaded connections can lead to a considerable leakage of CWA under conditions close to normal. If heating is several tens of degrees higher than the storage temperature (T_{st}), thermal-mechanical stresses due to the thermal expansion of the liquid CWA can be high enough in the sealed metal vessel, which is practically

V.M. Kolodkin and W. Ruck (eds.),
Ecological Risks Associated with the Destruction of Chemical Weapons, 247–252.

completely filled, that the munitions casing will be deformed considerably or even destroyed.

The goal of this work is a deterministic assessment of the possibility of maintaining the sealed state and the safety margin of the munitions casing in the conditions of real thermal effects.

2. SIMULATED PRESENTATIONS OF THE OBJECTS AND PROCESSES UNDER STUDY

2.1 Munitions model

A simplified munitions model, that is, two axis-symmetric coaxial cylinders, was selected for performing the assessments. The space between the cylinders is filled with chemical agent (CA). The external cylinder has internal and external radii R_1 and R_2 and a total length L. The same dimensions of the internal cylinder are r_1, r_2 and l. It is assumed that $L>>R_2$ and $l>>r_2$. The external cylinder has an intake closed with a plug with threaded connection and a sealing lead gasket. The internal cylinder is embedded into the external one on the contact surface on the face plane and closed with a metal plug with threaded connection. All the munitions parts are made from high-strength steel.

2.2 Physical mechanisms of increase of the intrinsic pressure in the munitions at the increase of the temperature of the ambient air

Considerable variations of the ambient air temperature are possible during storage or transportation of the munitions to the destruction site. The causes of such variations can be associated with usual changes in the weather conditions (the range of temperature variations from the sun heating in summer can reach 60^0C and higher) or accident situations (in the case of fire this range can reach 2000^0C). In all such cases a new mechanism of CWA leakage can appear, caused by the high value of the coefficient of thermal expansion (α) of liquid hydrocarbons. In the case of complete filling of the munitions with liquid CWA the temperature change ΔT would lead to a volumetric deformation, given by

$$\Delta V/V \approx \alpha \Delta T. \qquad (1)$$

This will increase the pressure in CWA

$$\Delta p = \frac{\alpha}{\gamma} \Delta T, \qquad (2)$$

where γ is the liquid compressibility. Thus, a temperature change of only 10°C would lead to the increase of the intrinsic pressure to $2 \cdot 10^{7}$Pa $\approx$200 atm. In these conditions the plastic flow of the casing and the munitions destruction can occur. The manufacturing methods provide for a 95% filling of the munitions. This means that, when heated to 50°C above the storage temperature, all the free internal volume will be filled and further temperature increase will lead to an irreversible disastrous deformation of the munitions.

2.3 Plastic deformation and destruction of munitions at heating

The conventional scheme of plastic deformation of the munitions by the pressure in the space filled with CA can be presented as follows. The degree of deformation of both cylinders is maximal in the center and minimal on the edges due to the rigidity of the ends. Assume the plastic deformation of the cylinders on the ends equal to zero and the deformation value in the middle will be evaluated like in the model of an infinitely long thick-wall pipe. The pressure in the material of the internal cylinder is as follows:

$$p_1 = \frac{2\sigma_T}{\sqrt{3}} \ln \frac{r_2}{r_1}, \qquad (3)$$

where σ_T is the limit of flow of the pipe material under strain. The pressure in the material of the external cylinder can be described with the relation given below:

$$p_2 = \frac{2\sigma_T}{\sqrt{3}} \ln \frac{R_2}{R_1}, \qquad (4)$$

which can be applied to the analysis of the stressed state of the external cylinder. If both cylinders are made from the same material, the cylinder that will begin deforming first will be the one with the lowest radii ratio. If this ratio would be equal for all munitions, it would be possible to predict their behavior exactly. However the exact description is impossible due to possible deviations in the production process. Three variants in which the munition's behavior differs considerably are admissible in principle. The first variant is:

$$(R_2 - R_1)/R_1 \le (r_2 - r_1)/r_1 .$$

For the 122-mm shell it can be realized by the set: R_2=61 mm, R_1=51 mm, r_2=24 mm, r_1=20 mm. Here $(R_2-R_1)/R_1$=0.196 $<(r_2-r_1)/r_1$=0.2. The second variant is:

$$(R_2 - R_1)/R_1 \ge (r_2 - r_1)/r_1 .$$

If, for example, R_2=61 mm, R_1=50 mm, r_2=24 mm, r_1=20 mm, then $(R_2-R_1)/R_1$=0.22 and this exceeds the ratio $(r_2-r_1)/r_1$=0.2 slightly. The third variant provides for a considerable excess of the ratio $(R_2-R_1)/R_1$ over $(r_2-r_1)/r_1$. For example, in the case of R_2=61 mm, R_1=46 mm, r_2=24 mm, r_1=20 mm, then $(R_2-R_1)/R_1$=0.33, which differs considerably from the value $(r_2-r_1)/r_1$=0.2 from the point of view of the theory of destruction.

Consider the process of plastic deformation and loss of integrity of munitions in the first variant. For steels $\sigma_T \approx 2\cdot 10^8$ Pa, hence, for the beginning of the plastic flow of the external cylinder the intrinsic pressure of CWA would need to be increased to the value $p_2 \approx 4\cdot 10^7$ Pa. This would occur with the increase of the temperature of the ambient air to 70°C higher than T_{st}; the value of which, $T_{pl} = T_{st} + 70^0 C$, is denoted the temperature of the onset of the plastic flow of the 122-mm shell.

The development of the plastic deformation will lead to the fall of the intrinsic pressure to the value of δp due to the increase of the volume between the cylinders:

$$\delta p \approx 2\delta R_1 / 3\gamma R_1 , \qquad (5)$$

where δR_1 is the change of the radius during the plastic deformation. The radius r_1 does not change in this case. To compensate for this fall by heating, the temperature of the ambient air would need to be increased by the value

$$\delta T \approx 2\delta R_1 / 3\alpha R_1 . \qquad (6)$$

Hence, as the temperature increases the radii R_1 and R_2 will grow continuously until the casing fails.

At the cold deformation of steels ($T \leq 0.4T_{area}$) the destruction takes place if the deformation degree $\varepsilon \approx \delta R / R$ exceeds the values (0.15÷0.56). However, even at the value $\varepsilon \approx h_p / r_p$ (where h_p and r_p are the height and the radius of the intake thread, respectively) an almost complete loss of integrity of the munitions can occur. In practice, the ratio of the thread height to the diameter of the hole normally does not exceed 0.1. Therefore, at $\varepsilon \approx 0.2$ the diameter of the intake will increase so that the thread cannot hold the plug. As the intrinsic pressure is considerable, forcing the plug out of the intake will happen earlier, approximately at $\varepsilon \leq 0.1$. Assuming $\varepsilon \approx 0.1$, we will get the temperature of the complete loss of containment of the 122-mm shell:

$$T_f = T_{st} + 130^0 C . \qquad (7)$$

The external diameter of the munitions can change by that moment from 122 mm to the values of about (140-150) mm and the internal cylinder will be in the initial undistorted state.

Let us notice in addition that at $T = T_{pl}$ the intake will be already partly unsealed. Though the thread will keep the plug in the hole, the liquid CWA will escape outside under the high pressure through the widened gaps in the threaded connection. Therefore, T_{pl} should be considered the real temperature of munitions failure.

The similar consideration is possible for the other variants.

3. CONCLUSIONS

1. During storage and transportation of the CA-filled munitions the environmental damage associated with the CA leakage into the

environment due to the plastic deformation of the munitions casing at the temperature variations of the ambient air are possible.

2. For the 122-mm munitions at the temperature increase to $T_{pl} = T_{st} + 70^0 C$ the possibility exists of a new CWA leakage mechanism associated with the loss of integrity of the threaded connection of the intake due to the plastic deformation of the external cylinder. The process occurs under the condition that $(R_2 - R_1)/R_1 \leq (r_2 - r_1)/r_1$. Further temperature increase to $T_f = T_{st} + 130^0 C$ leads to a complete loss of integrity of the intake. The destruction of the external cylinder in other points is less probable.
3. When $(R_2 - R_1)/R_1$ considerably exceeds $(r_2 - r_1)/r_1$, the mechanism of failure due to the plastic deformation and destruction of the internal cylinder becomes possible. The external cylinder remains in the initial state. To realize this mechanism, the temperature of heating should be $T_c = T_{st} + 90^0 C$.
4. At $(R_2 - R_1)/R_1 \geq (r_2 - r_1)/r_1$ the CWA leakage mechanism is possible at the simultaneous deformation of the both cylinders. It always ends by the loss of integrity of the intake of the external cylinder.

ECOLOGICAL RISKS AND ECOLOGICAL INSURANCE IN RUSSIA (IN CONNECTION WITH THE PROBLEM OF CHEMICAL WEAPONS DESTRUCTION)

Vladimir G. Gorsky[1]
[1]*Russian Research Center (GOSNIIOKHT), Moscow (Russia)*

Abstract: This work considers issues of environmental risk and insurance in Russia.

Key words: environmental risk, prediction

On November 7, 1997, Russia ratified «The Convention on Prohibition of Development, Manufacture, Accumulation and Application of Chemical Weapons and their Destruction». The Russian Federation has assumed liability to destroy stocks of chemical weapons within 10 years with possible five years' prolongation of the destruction time.

According to this obligation the Federal Law No. 76 from 2.05.1997 «On Destruction of Chemical Weapons» and the Federal Program «Destruction of Stocks of Chemical Weapons in the Russian Federation» has been accepted.

The chemical weapons are located on seven Russian arsenals:

- in the Bryansk region, town of Pochep
  - about 18.8 % of the stockpile
- in the Kirov region, settlement of Maradykovsky
  - about 17.4 % of the stockpile
- in the Kurgan region, town of Shchuch'ye
  . about 13.7 % of the stockpile
- in the Penza region, settlement of Leonidovka
  - about 17.2 % of the stockpile
- in the Saratov region, township of Gorny
  - about 2.9 % of the stockpile
- in the Udmurt Republic, town of Kambarka
  - about 15.9 % of the stockpile
- in the Udmurt Republic, township of Kizner
  - about 14.1 % of the stockpile.

The total amount of chemical weapons in Russia is 40,000 tons.

V.M. Kolodkin and W. Ruck (eds.),
Ecological Risks Associated with the Destruction of Chemical Weapons, 253–260.

By May 1, 2003, in the arsenal of Gorny 400 tons of yperite were destroyed. Intensive preparatory work for CWA destruction is conducted at other arsenals.

The operation of facilities constructed for destruction, transportation and storage of chemical weapons can result in pollution of the environment. Ecological insurance provides compensation for such damage as well as additional financing to increase ecological safety of similarly dangerous facilities.

The legal basis for regulation of ecological safety in the Russian Federation is the Constitution of the Russian Federation of 12.12.1993 and the Civil Code No. 146-ФЗ of 26.11.2001.

The following Russian Federation legislation form the modern legal basis of the maintenance system for ecological safety:

- «Act on the Destruction of Chemical Weapons», No. 76 of 2.05.1997;
- «Act on the Industrial Safety of Dangerous Industrial Plants», No. 116 of 21.07.1997;
- «Act on the Preservation of the Environment», No. 7 of 10.01.2002.
- Bases of ecological insurance are incorporated in the federal Acts:
- «On the Setting up of an Insurance Project for the Russian Federation», No. 204 of 20.11.1998;
- «The Insurance Act».

On the basis of this legislation, the typical regulations about voluntary ecological insurance are developed, but this form of insurance does not allow the active development of the sector of the insurance market. The Federal «Act on Ecological Insurance», providing for mandatory types of insurance, serves this purpose. However, at the present time, this bill is still under consideration by the State Duma. It limits possibilities for insurance of storage facilities, transportation and destruction of chemical weapons in the Russian Federation.

Ecological insurance in the Russian Federation is understood as insurance against the responsibility for damage to life, health and property of thirds, and the environment in case of emergency pollution. Therefore the term « Ecological Liability Insurance» is more suitable.

Ecological insurance is closely connected with the concept of ecological risk.

In a number of Russian and foreign publications [1] ecological risk is assumed as only the risk to environmental elements. Thus, risk to life and health of people is an independent category. To avoid different interpretations, the term "generalized ecological risk" (GER) is expedient to refer to ecological insurance. The generalized ecological risk is the predicted damage (measured by the frequency of emergency events) to the population and environmental elements from the emergency pollution generated by any industrial object.

The concept of ecological insurance cover needs a very precise description. In this question there is no unanimity either.

Prof. G.A. Motkin [2] is the founder and the ideologist of the ecological insurance in the Russian Federation. Now in the Russian Federation three basic approaches for the estimation of danger level of industrial facilities under ecological insurance cover have been established and are now applied.

The first approach, which can be called qualimetric, consists of the following. First a full set of n quantitative or qualitative individual parameters, which characterize the danger represented by a facility under discussion is selected. For each of them the corresponding scale of measurements is established. Then, with the help of a group of experts, the set of the similar facilities subject to insurance is analyzed. Each expert estimates the parameters of danger of facilities under discussion and derives a total estimation as a class of danger, expressed usually in points.

After that, two algorithms are used. According to one of them a certain generalized parameter of danger is determined, and a functional bond of the generalized parameter with individual parameters is established. The second algorithm consists of n-dimensional space splitting into the crossed areas corresponding to different classes of danger.

The second approach, stated in /2/, is based on the analysis of the retrospective data about volume and quality of harmful emissions and dumps from industrial facilities in the environment for a certain time interval. This approach proceeds from the «**polluter-pays principle** ». On the basis of this, analysis critical values of emissions (dumps), usually exceeding by several times the regulated values determined by maximum permissible emissions (dumps), are established. Actual excess of critical values of emissions (dumps) for the controllable period of time is qualified as emergency environmental contamination.

The annual damage after emergency environmental pollution is estimated as follows:

$$T = \sum_{j=1}^{J} V_j , \qquad (1)$$

where V_j is annual damage for the j-th recipient, in monetary units/3/.

Calculation of damage can be carried out, for example, according to /3/.

We note that within facilities insured accorsing to this approach are divided into classes depending on the level of excess of critical values of emissions (dumps).

It is necessary to note that, in the first and second approaches, a risk concept, strictly speaking, is not used.

The third approach is based on the modern theory of the accident risk generated by industrial facilities /4, 5/. The modern theory of accident risk analysis includes 4 basic stages: preliminary hazard identification (PHI); analysis and assessment of possible accident consequences (AC); frequency analysis of accidental events (FA); prognosis and evaluation of accidental risk (PER).

PHI is intended to identify the causes of accidents, to determine the scenarios of accidents, and to choose the most dangerous from them.

The purpose of **AC** is to determine possible damage from each of the most hazardous accidents, supposing that it will take place with probability of 100%.

FA is intended to estimate the frequency of occurrence of each most dangerous predicted accident (the frequency is measured in time-inverse units).

The purpose of **PER** is to predict the value of cumulative accident risk R determined generally as

$$R = \sum_{j=1}^{J} \sum_{i=1}^{I} F_i U_{ij} \ , \qquad (2)$$

where F_i - frequency of predicted accident i, $[F_i]$== [year^{-1}]; $i = 1,...,I$; U_{ij} - probable damage for the recipient j as a result of accident, in monetary units.

The management of emergency risk intended for development of recommendations for reduction of probable damage and frequency of predicted accidents with the purpose of achievement of a comprehensible level of accident risk is carried out iteratively.

The block diagram of the accident risk analysis is presented in the following figure:

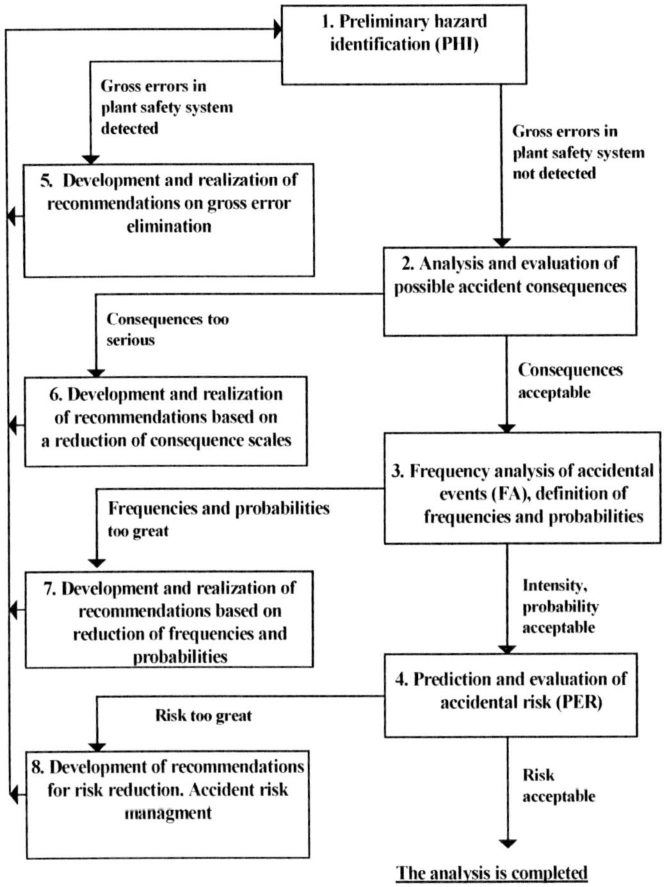

Fig.1 The block diagram of accident risk analysis

It is necessary to note, that specific techniques of ecological insurance can contain the elements borrowed from the three approaches mentioned above.

With regard to ecological insurance of storage facilities (CWSO), transportation (CWTO) and destructions (CWDP) of chemical weapons, it is necessary to consider their specific features.

1. The number of the facilities connected with chemical weapons is rather limited in comparison with the total number of chemically dangerous facilities. Moreover, in the Russian Federation, the construction of

CWDP is only being started . Additionally, they will not be of the same type.

2. CWSO and CWDP, as a rule, are place in locations remote from other industrial facilities, i.e., sources of environmental pollution.
3. As for the facilities under consideration as pollutants, it is necessary to take into account not only CWA, but also other substances used in their destruction. It is especially essential when recipients are elements of the environment.

By virtue of the above-mentioned features of CWSO, CWTO and CWDP, it is hardly probable that their insurance will estimate the level of danger within the first and second categories. It is expedient to take the third approach as a basis. Especially, taking into account that all facilities connected with the destruction of chemical weapons according to the «Act on the Industrial Safety of Dangerous Industrial Facilities» are obliged to have the Declaration of Safety, in which the estimation of the level of danger of the facility is carried out on the basis of the modern theory of accident risk analysis.

However there are a lot of difficulties to carrying out such an approach. First of all there is only a weak theoretical study of emergency pollution on the scale concerning environmental elements (atmosphere, soil, surface waters) with accounting of mass transfer between them, persistency of substances, and long-range distribution of substances. Here it is necessary to specify the difficulties in forecasting the toxic effect for biotic elements, i.e., representatives of flora and fauna.

Therefore it is expedient to form the ecological insurance of the subject facilities on the bases of combination of the first and second approaches to estimate the object danger level. In this case, the value of "generalized ecological risk"(R_t) must be calculated as follows:

$$R_t = \sum_{j=1}^{J_1} \sum_{i=1}^{I} F_i U_{ij} + \sum_{j=J_1+1}^{J} V_j \,, \qquad (3)$$

where J_1 - number of recipients, for which the possible damage is predicted, with the help of accident risk theory, $J - J_1$ - number of recipients for which the damage is determined within the second approach.

In GER estimation it is important not only to give the quantitative risk estimation, but also to carry out the analysis of its uncertainty [6]. In this connection modern methods of the interval analysis can be used. The example is found in [7].

Finally, we think that, at such a representative forum as this meeting, it would be useful to discuss the question of accounting for ecological insurance not only for the damage after emergency environmental pollution, but also after regular emission of industrial facilities, namely CWSO, CWTO and CDWP.

References

[1] Onishchenko, G.G., Novikov, S.M., Rahmanin, J.A., Avaliani, S.L., Bustueva, K.A. (2002) Basic Principles of Health Risk Assessment at Influence of Chemical Substances Pollution of the Environment. Moscow, pp 408.

[2] Motkin, G.A. (1996) Bases of Ecological Insurance. Moscow, pp192.

[3] Methodical Recommendations According to Damage from Accidents on Dangerous Industrial Facilities. (2002). RD 03-496-02, Moscow.

[4] Guidelines for Chemical Process Quantitative Risk Analysis, Center for Chemical Process Safety of American Institute of Chemical Engineers. (1989). N.-Y., 585 p.

[5] Gorsky, V.G., Shvetzova-Shilovskaya, T.N., Kirsanov, V.V., Tereshenko, G.F. (2002). Analysis of Accident Risk and Safety of Chemical Dangerous Facilities. Chemical Industry (in Russian) 4: 3-16.

[6] Gluhan, E.N., Chernyshova, M.P., Gorskij, V.G., Shvetzova-Shilovskaya, T.N. (2003). Analysis of Uncertainty at Risk Assessment of Chemical Plants. Management and Information Technologies. Proceedings: All-Russia Scientific Conference, St.-Petersburg, Russia, Vol 2, pp 252-257.

[7] Gorsky, V., Shvetzova-Shilovskaya, T., Voschnin, A. (2000). Risk Assessment of Accidents Involving Environmental High-toxicity Substances. Journal of Hazardous Materials Vol 78, 1-3: 173-190.

INSURANCE OF RISK OF ENVIRONMENTAL CONTAMINATION DURING DESTRUCTION OF CHEMICAL WEAPONS

Gennadiy Motkin
Head of laboratory at the Institute for Market Economy Problems of the Russian Academy of Sciences (IPR RAN), Moscow (Russia)

Abstract: This work considers various approaches to insurance of environmental risks.

Key words: environmental risk, insurance

A definition of environmental insurance should be made on the basis of characterizing the specific features inherent both in environment (due to hazard impact) and in property and liability insurance.

This will help us to explain the difference in understanding of the notion of environmental insurance by Russian and Western experts and, thereby, to determine its role and place in the formation of a national market economy.

First of all, accidental environment pollution insurance is oriented towards risks with little-known origin, and is thus difficult to evaluate and quantify adequately. If we consider risk as "mathematical expectation of loss function while defining the parameters of mathematical model and of its structure"[1], which is identical to definition of risk[2], then its amount depends upon, at least, the following extremely important components: volume of hazardous substance emission, type of recipient, exposure period, and time and rate of environmental hazard of this chemical or physical element. The creation of an integral index of accidental environment pollution consequences, which adequately identifies the level of economic losses, is hardly feasible. But it seems that it is not necessary to make one. One should create the methods of evaluation of loss caused by accidental environment pollution, as recognized by users and by insurance and insurers in our case.

A specific feature of accidental pollution is the impossibility to compare its consequences with the negative impact of so-called constant anthropogenic pressure on nature. Meanwhile the constant hazard emission into environment in quantities exceeding temporarily admissible levels may be qualified as accidental pollution taking into account its negative consequences. This gives a foundation for discussing quantitative and qualitative techniques to identify accidental environment pollution [3]. However it is impossible to calculate the probability of a situation when

V.M. Kolodkin and W. Ruck (eds.),
Ecological Risks Associated with the Destruction of Chemical Weapons, 261–266.

environmental impact could be definitely framed, on basis of current data bases. There is no information on accidents with registered environmental impacts, or the access to this information is still not available. This fact is closely connected with the absence of a strict notion of an environmental accident. One can cite many examples of accidents and technogenic catastrophes, even their frequencies, but there are no methods of environmental hazard assessment for this or that type of production activity that meets the requirements of environmental insurance. Insurance environment auditing should become the core of the environmental safety assessment of different enterprises and businesses. It should find the answers to only two very important questions:

- what is the probability of an environmental accident in a specific facility, including the system of environmental insurance? and
- what is the value of losses that recipients could incur as a result of the environmental accident?

There are several methodical approaches to the problem of insurance environment auditing that are considered now.

1. Industrial production hazard is identified, using a list of hazardous chemical substances, used in least admissible quantities.
2. Industrial production hazard is identified by analyzing the rates of exceeding least admissible standards of environment impact.
3. Industrial production hazard is defined by calculating pollution risk values and caused assessed damage.

These enumerated "loss creating" characteristics of emissions of hazardous substances lead to specific requirements for methods of environment pollution risk insurance.

The subject being discussed is losses, caused by **accidental environment pollution**. That means that in environmental insurance we consider the losses due to emission from one source of a certain amount (in accidental volume) of hazardous substance and its negative effects on certain recipients. There is identification of the guilty and recipient in the accidental environment pollution liability insurance. The share of the polluter is not identified in property insurance. This means that financial coverage of insurance premiums is ensured not only by the different sources, but received insurance payments are used for different purposes by insurer.

Environmental insurance in the form of accidental environment pollution liability insurance is aimed at environmental safety guarantee and the third party damage reimbursement (of course, taking into account the commercial benefit of insurance and insured).

Environmental insurance in the form of property liability can't be considered as component of environment safety guarantee and is aimed only at the of compensation the loss incurred by the insured.

In this aspect environmental insurance differs from the other types of insurance activities, medical for example, though it covers the same types of parties, called third-party insurance. Losses, expressed for example in population health deterioration, are defined in environmental insurance, using quite different principles than, for example, medical. It is necessary for environmental insurance to identify the pollution sources and recipients with maximum reliability, and utilize this information to define price and compensation policy. Medical insurance uses other assumptions, fully valid, but quite different: each enterprise, paying salaries to its employees, bears a certain financial burden: compensation of losses due to population morbidity, irrespective of whether it –the enterprise– is the source of accidental pollution or not.

The definition of environmental insurance as a liability insurance of enterprises, i.e., sources of high environmental danger and property interests of insurants in case of accidental (sudden, unpremeditated) environmental pollution, ensuring the compensation of environmental pollution losses and creating additional sources for financing of environment protection measures [4], is based upon the foundations described above. Its main goal is additional financial support of environmental safety, taking into account all interested parties: the insured, insurances and third parties.

A very wide range of different regulations and other documents was elaborated for property insurance activity. Where environmental pollution liability insurance is concerned, it should find a special sector in the methodical and legislative field of insurance.

It is clear that the problem of finding the new sources for the financing of environmental protection measures is so acute that only private capital has additional financial reserves in the present situation. To find attractive directions for private capital investments is one more environmental insurance function as we see it.

There is the point of view that the adoption of the federal "Act on Environmental Insurance" will result in enterprises, which pollute, becoming involved in the environmental insurance process. But liability without economic interest or incentive is useless. This act must be integrated into economic relations, and take into account the current situation in insurance business and the environment protection activities of the insured.

At present, there are several acts and regulations defining the framework of environment pollution liability and the insurance role in this sphere.

An article in the "Environment Protection Act" directly specifies that, in the Russian Federation, environmental insurance of enterprises, organizations and people, their property and income, takes place in the case of environment and natural hazards, accidents and catastrophes. So, insurance is not only the source of potential income, but also an activity for prevention, elimination and compensation of harm (in environmental economics we use the notion of damage, in legal practice - losses), caused to victims. Economic damage due to environmental pollution is the sum of costs needed to prevent environment pollution impact on recipients (when partial or complete prevention is technically feasible) and of environmental pollution losses[5]. In the Civil Code, it is specified that "losses are the costs that third party has incurred or has to bear in order to restore its violated rights, losses or property damage (real damage) as well as income that would be received in typical economic conditions is this right had not been violated (income loss). If the third party, having violated this right gained some benefit, the other party, whose right was violated, has right to demand the compensation of income loss in amount not less than these benefits in addition to the compensation for the other losses."[6]

So, theoretically, insurance premium value includes two components. The first one is accidental pollution prevention cost. This is additional unreasonable cost incurred by the insured during the insurance contract lifetime in the case of absence of accident. This cost is a component of potential losses from the point of view of society and of the third party that benefits from the accident environment pollution liability insurance contract. The insurance, realizing that and evaluating potential insurance compensation, either allocates financial means for accident prevention themselves, or obliges the insured to take environmental protection measures. These measures either can either be fulfilled or taken into account in the calculation of the insurance premium.

The second component of the insurance premium is the assessment of pollution impact on recipients. These losses are due to the impact on the recipient from emissions of harmful substances. Unlike the first type of losses, a third party is also involved. Environmental insurance is liability insurance of accidental environment pollution caused by sources of high hazard in both cases.

Not only recipients, the third party who benefits from liability insurance, have accident pollution losses but also insurants themselves, who are at the same time recipients as well. Those who are neither recipients, nor polluters also can be insurants.

In connection with this we previously talked about insurer's policy differentiation. So, the insurer, while reimbursing the losses of the insured –

accident environmental polluter, doesn't create the incentives for the insured to prevent this pollution.

The insurer, while compensating the third party losses, liberates the insured polluter from elimination of accident consequences and prevention of future possible accidents. Insurance premiums are of particular importance in the insurer's behavior supervision. They cannot be unique either for industries or for selected insured enterprises. The same refers to liability limits for environment pollution risks accepted by the insurer.

Theoretical aspects of relationships between insurers and insured require possible case study simulation and elaboration of consequent methodical foundations.

Those who minimize future risks and social costs, benefit from the insurance process. As a result, market economy mechanism can be used as an instrument for risk management with possibility of considerable decrease in environmental damage. The use in practice of this direct economic incentive could be an effective addition to traditional methods of economic and legal management of relationships between society and nature. While analyzing foreign or national experience and materials on environmental insurance, one can identify four groups of fundamental problems of environmental insurance development.

The first one defines the essence, role and place of environmental insurance in real economy and considers this type of insurance activity to be of state importance as a component of ensuring the safety of national environment. This factor is the basis of conception for compulsory environmental insurance introduction with all its consequences (compulsory list of environment hazard enterprises, insurance premiums, included in product cost or in pollution charges, etc.).

The second group comprises the principal statements of insurance environmental auditing, that gives possibility to solve problems of defining objects of insurance field (to evaluate environment hazard rate of the enterprises and businesses, to calculate the possible loss, etc.).

The next group of problems forms the legal space for environmental insurance. In Russia there is a real possibility to create a comprehensive legal base for environmental insurance development by adopting the Federal "Act on Environmental Insurance" and to ensure its effectiveness by elaborating other consequent regulations that constitute the fourth group of fundamental problems of environmental insurance development.

References

[1] G. Gorsky, G.A. Motkin, T.N. Shvetzova-Shilovskaya, V.K. Kourotchkin What is the risk? (1995). In collected articles "Works of the First All-Russia Conference 'Theory and Practice of Environmental Insurance' ". - Moscow

[2] S.A. Aivazyan, J. N. Tjourin (1990) Finansy i Statistica: Reference Book for Applied Statistics, Translated from English.// Red. - v.2. - Moscow,

[3] G.A. Motkin. Methods of Premium Definition in Ecological Insurance system (1995) // "Economics and Mathematical Methods", v.31, issue 1;
Моткин Г.А., Тулупов А.С. (отв. редакторы). Развитие экологического страхования в Республике Башкортостан (2002) - ИПР РАН, Москва – Уфа

[4] G.A. Motkin. Foundations of Ecological Insurance (1996) - Moscow, "Nauka",.

[5] Temporary Typical Methods of Definition of Cost Effectiveness for Environment Protection Measures and Evaluation of Economic Damage, Caused to Economy by Environment Pollution (1986) - Moscow, "Economica"

[6] Civil Code of Russia (1996) - Moscow

4

SESSION III: PREVENTION

ABOUT SOME RISK ASSESSMENT PROBLEMS ASSOCIATED WITH HAZARDOUS FACILITIES EXPLOITATION

Anatoliy Michailov[1], Stanislav Petrin[1], Lada Petrina[1]
[1]*Russian Federal Nuclear Center– All-Russian Research Institute of Experimental Physics, (RFNC-VNIIEF), Sarov (Russia)*

Abstract: The desire to ensure an acceptable risk level in handling toxic substances leads to the necessity of developing safety criteria for the case of an accident while working with such substances. The solution of this task on the basis of the acceptable risk conception is connected with the need of a series of solutions to problems, which are summarized in the report.

Key words: risk assessment problems, safety criteria

Human industrial activity leads to the violation of natural ecological systems, as well as to the risk of loss of the balance of social and economic tasks, and preservation of a favorable environment with biological variety and natural resources to meet the needs of present and future generations. The urge to manage risk with the purposes of stable society development leads to the necessity of risk assessment by methods of probability safety analysis (PSA), selecting problems which cause special concern and possible ways of their solution. The performance of PSA provides examination of all sources of danger characteristic of considered hazard industrial facility (HIF), and modes of HIF influence on the environment with normal operation and in accidents according to every possible scenario.

In RFNC-VNIIF for over 10 years HIF operation safety work analyses have been performed where work with nuclear, radioactive, explosive, and toxic hazard substances has been carried out. A common analysis strategy applied is presented in Fig. 1. It is represented in an iterative loop at the beginning of which the initial information about HIF and purposes chosen for the analysis safety criteria are given; and at the end – an inference about the possibility of the chosen criteria used.

V.M. Kolodkin and W. Ruck (eds.),
Ecological Risks Associated with the Destruction of Chemical Weapons, 269–278.

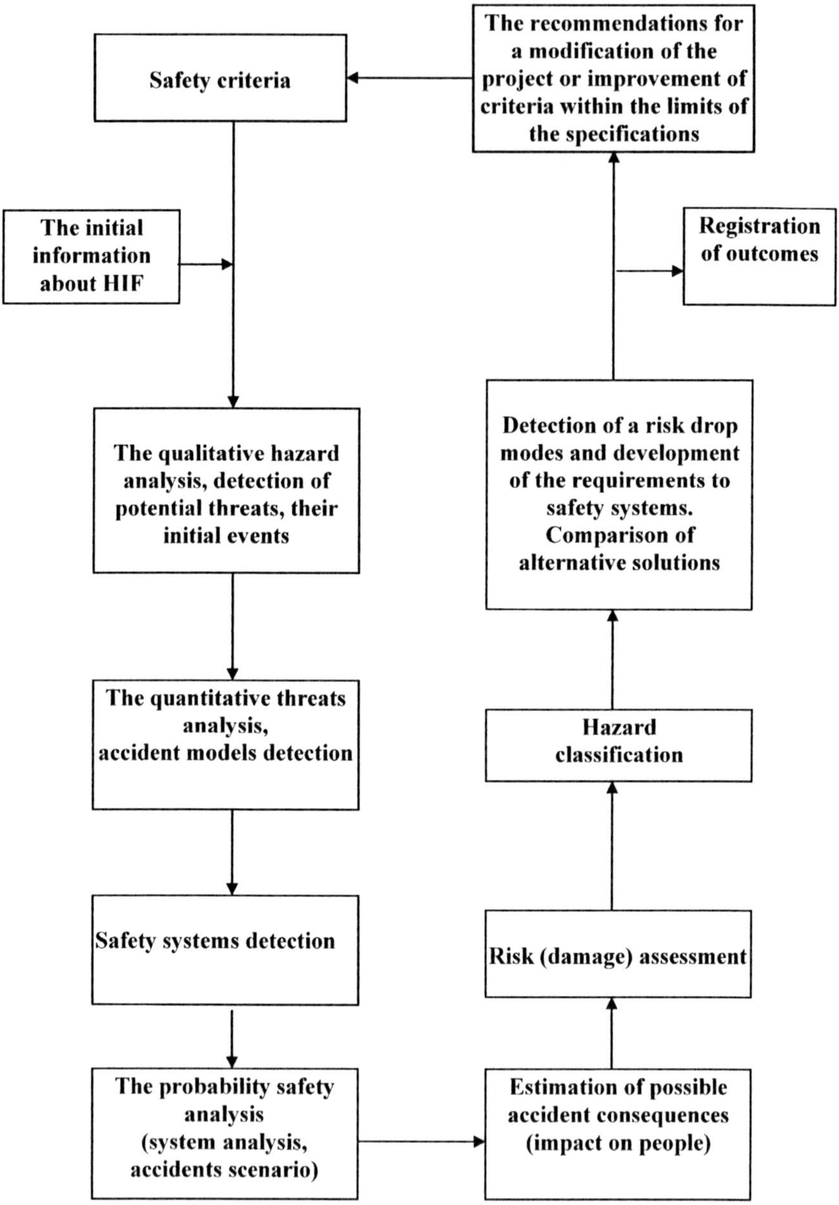

Fig. 1: Common strategy of the analysis

In the analysis strategy there are three distinguishing basic components:

- safety criteria choice. As a basis any of the accepted concepts and also operating normative documentation are used;
- the qualitative and quantitative hazard analysis. The procedures PHA, HAZOP, WHAT IF, FMEA [1] are used at performance of the qualitative analysis. They allow revelation of a wide spectrum of potentially dangerous threats. In the quantitative hazard analysis deterministic estimation methods are used, which allow a selection of models of accidents for the occurrence of which the intensity of effects at the origin of initial events (realization of threats) is sufficient;
- the probability safety analysis (including safety-systems analysis, their reliability, effectiveness of function under various conditions, and also estimation of possible accident consequence analysis). At this stage risk is estimated, basic possibility of safety criteria observance is established, danger is classified, risk cut sets are discovered, requirements for safety systems are established, and alternative solutions providing HIF ecological safety are compared.

One of the major analysis stages is the choice of safety criteria, when the inference about an acceptability of available risk or necessity to reduce its level can be drawn. Here risk is understood as probability of causing harm to citizens' life or health, persons' or officials' property, state or municipal property, environment, and life or health of animals and plants, taking into account the degree of this harm. It is expressed as probable material damage connected with necessary compensation to the injured (citizens, persons and officials, organizations) in accidents and also expenses for rehabilitation of the environment. At the normal operation of the HIF, if its negative impact on people and environment is above the acceptable level, it should be excluded at the expense of the safety criteria choice and protective measures.

It is necessary to note that, until recently toxicity safety criteria (in cases of accidents in particular) have been given little attention in Russia. Work with toxic substances applied criteria that took into account only their chronic inflows to the human body and environment.

Human beings, as part of the ecological system, are usually the most sensitive (critical) element. It has been proved experimentally for cases of radiation impacts and, as a rule, for toxic impacts. Therefore at first approximation the purpose of the safety criteria choice can be preservation of people's health both with normal HIP operation, and in industrial accidents connected with individuals' errors, equipment failures, and technical and natural impacts on HIF. Health preservation means absence of diseases leading to human death, or absence of occurrence of hereditary deviations in the subsequent generations. For the purposes for HIF safety analysis we, as a rule, use sanitary - hygienic safety criteria based on the acceptable risk conception. It corresponds to the modern approach to the hazard analysis, in which the addition of a new HIF does not significantly increase the individual and social risk level; it does not lead to unacceptable direct and indirect impact on people.

The acceptable risk concept consists of the following [2-4]:

- mankind intuitively accepts those aspects of practical activity, where (individual maximum death risk at the moment of accident or due to illnesses initiated by accident or regular emission of dangerous substances) does not exceed a low value $\sim 10^{-5}$ *1/year*, if this risk is not compensated. Such risk is at a level of less than 0.1 % of natural death risk. Occupations which increase individuals death risk should be prohibited;
- the personnel of the industrial facilities accepts those aspects of industrial activity, where the individual death risk is in a range from 10^{-2} to 10^{-5} *1/year*, and the acceptable average level of death risk during the industrial activity in the developed countries is equal to $2.5 \cdot 10^{-4}$ *1/year*. A higher level of risk is financially compensated. It is supposed that, by working with dangerous substances, the HIF personnel will be exposed to risk of not more than indicated average value, and the auxiliary staff working within the HIF exposure zone limits, will be exposed to 10 times less risk;
- the social risk, which is an integrated averaged individual death risk in all possible accidents, is considered acceptable, if it does not exceed 10^{-4} *individuals/year*; as for exposure of child's institutions or/and boarding schools, the social risk must not exceed 10^{-5} *individuals/year*. The social risk is defined as the sum of the number of possible victims, i.e. deaths and significant hereditary deviations, in separate accidents, based on their probable frequencies.

Let us illustrate our approaches to the safety criteria choice while working with compounds of uranium - 238, which is a toxic substance and is sometimes used for military purposes (not long ago the USA used depleted uranium during the war in the Balkans).

The analysis of the published information about uranium compounds toxicity (according to the experimental data on tested animals with their recalculation for the human factors of Y. Rybolovlev [6]) gives a picture presented in the table 1.

The safety criteria to work with uranium, worked out in accordance with the concept of acceptable risk and radiation danger of uranium at chronic inflow to the human body, are submitted in the table 2. According to these criteria in normal HIF operation the established restrictions on TLW for the HIF personnel (persons of group À) are kept, though accumulation of uranium in the human body is at the threshold sensitivity level when the state of the human body is stressed and some individuals may be prone to disease development. The risk to personnel is above average, but the personnel of group A receive the appropriate compensation. In the future there is a possibility to reduce the TLW value.

Let's note that in accidents (one-time impacts) the danger of uranium getting into the human body is defined by its chemical toxicity, not by its radiation danger. One-time inflow of uranium to the human body of the group A personnel during planned accidents is chosen in a way so that there are no acute diseases in accidents and residual amounts of uranium in a human body (after removing in a short period of time 90% of its soluble or 40% low-solubility compounds) do not cause diseases later in time.

It is suggested that the inflow of uranium compounds to the human body during planned accidents be limited so that there are no acute diseases right after the accident. In non-planned accidents the accumulation of uranium in lungs should not cause malignant formations, nor the necessity to evacuate people from the inhabited zone to lower the toxic level.

The offered criteria can be applied in the following ways:

- to estimate integral dangerous impacts on people it is necessary to proceed from the assumptions which provide establishment of criteria with probability not less than 0.95-0.99 at the expense of the level of confidence in estimations of expected frequency of accidents and use of conservative suppositions while estimating dispersion of an aerosol uranium emission cloud in the environment at the time of emergencies. In calculations of dangerous impacts the supposition is made of the worst meteorological conditions (direction and velocity of wind, condition of atmospheric stability) and absence of means of individual protection of the respiratory system of people belonging to B and C groups;
- to estimate dangerous impacts for the people outside of a HIF exposure zone, maximum (not average) doses for individuals who are on the track axis of the emission cloud on the boundary of the HIF exposure zone and inhabited zone should be taken into consideration;
- for all categories of people, only the inhalation inflow of uranium to human's body at the moment of the cloud passing is taken into account. It also refers to people residing in the polluted area during accidents.

Table 1: Indices describing danger of chronic and single inflow of compounds of uranium to human's body through inhalation

	Critical organ	At chronic impact		At one-time impact	
		Accumulation in body *mg*	Consequences	Dose *Mg*	Consequences
Low solubility uranium compounds $(NH_4)_2U_2O_7$, UO_2, U_3O_8, UF_4	Lungs	< 11	Almost absent	25	Acute disease
		9-12.7	Minimum operating accumulation	150	LD_{50}
		50 *	Malignant cancer, stipulated by chemical toxicity		
		100-150	Acute disease		
		246	Malignant cancer, stipulated by radiation		
		570	LD_{50} –50% of human's death		
		1000	LD_{100} –100% of human's death		
Soluble uranium compounds UF_6, $(NO_3)_2UO_2$, UO_2F_2	Kidneys	< 1.2	Almost absent	0.9	Threshold sensitivity
		1.2	Threshold sensitivity	9	Acute disease
		5	Moderate diseases	60	LD_{50}
		10-50	Acute disease	80	LD_{100}
		190	LD_{50}		

* A hypothetical index, since there is no authentic experimental information.

Table 2: Planned sanitary - hygienic levels to work with low dissolved uranium compounds (in brackets - with fast dissolved uranium compounds)

Situation		Maximum admissible levels of uranium inflow to a human body for			Death risk at toxic impacts at the expense of inflow of uranium to a human body, *1/year*		
		HIF personnel (A)	personnel of the next buildings, sites in HIF exposure-zone (B)	individuals from an inhabited zone and personnel of the next sites, buildings outside of a HIF exposure-zone (C)	A	B	C
Normal conditions		*LAI_A = 220 *mg* ^{238}U (550 *mg*)	LAI_B= 70 *mg* ^{238}U (160 *mg*)	LAI_C= 25 *mg* ^{238}U (25 *mg*)	$5 \cdot 10^{-4}$	10^{-4}	10^{-5}
Planned accidents	Estimated probable frequency of accident, *1/year*	Single inflow of low solubility ^{238}U compounds to a human body during accidents, *mg* (on uranium)			Integrated risk for 50 years		
	$P > 10^{-2}$	6 (3)	2 (1)	2 (1)	$2.5 \cdot 10^{-4}$	10^{-4}	10^{-5}
	$10^{-2} > P > 10^{-7}$	25 (9)	8 (3)				
Non-planned accidents	Maximum non-planned accident	Is not limited		For inhabitants of a residential zone single inflow of uranium to a human body during accident not more than 25 (9) *mg*	Is not limited		10^{-4}

* LAI- limiting annual inflow in body

The use of the similar approaches to work with poisonous warfare substances will provide a high level of safety for people; it follows the modern tendencies of development of the approaches to estimate safety of equipment using toxic substances and can satisfy public demands.

However the application of the described conception to develop ecological criteria of HIF safety connected with processing, storage and transportation of toxic (poisonous) substances, is connected with necessity of a series of solutions to problems. These problems are the following:

- it is necessary to establish whether there is a correspondence between levels of extreme admissible concentration of dangerous substances in the air of a working zone (TLW), daily average extreme admissible concentration of dangerous substances in the air of an inhabited zone and the level of acceptable risk. In Russia the methodology of the sanitary - hygienic standards establishment on the basis of the estimation of health risk has begun to take root only recently [5]. To solve this problem it is necessary to experimentally (which is very difficult) discover the threshold levels of impacts, when the lower level indicates absence of health damage or presence of life span increase;
- it is necessary to make sure that a human is a critical element of ecological systems (here a human is meant to have possibility to adapt to environmental changes when biological variety can vary due to HIF exploitation);
- it is necessary to establish a difference between mechanisms of impact, on people and the environment, of emergency emissions of toxic substances and regular small emissions of these substances (taking into account modification of their structure and synergism of aggregate substances impact) to make necessary corrections to values of admissible one-time emissions.
- it is necessary to define the rate of removal of dangerous substances from the human body and environment to establish the necessity of interference for the purposes of decreasing risk for people and rehabilitation of the environment;
- as for planned accidents, i.e. accidents happening with probable frequency of more than 10^{-7} *1/year*, it is necessary to establish values of admissible single inflows of dangerous substances to the human body (HIF personnel and population) by inhalation and through skin. The inflows that get into a human body with eating and drinking are supposed to be under control. At the same time the maximum individual and average social risk for people should remain at the acceptable level at any scenario of accidents. Both risks, the one of impact while the cloud of dangerous emissions is passing and the one

connected with residence on the polluted territory, should be taken into account. In this connection the level of admissible contamination of territory with dangerous substances should be regulated; this will reduce dangerous impact on people and modification of the environment (at the acceptable level of expenditures for its rehabilitation after accidents);

- it is necessary to establish admissible values at non-planned accidents of single inflows of dangerous substances into human's body (by inhalation and through skin). The individual risk for individuals should remain at the level of acceptable risk for the HIF personnel. Social risk (average over all possible variants of accidents) should be kept at the acceptable level. In this connection the level of admissible contamination of territory with dangerous substances should be regulated during non-planned accidents.

There are also problems of the choice of the acceptable level of material damage owing to accidents (for example, explosions of chemical substances).

The solution of the enumerated problems connected with storage, transportation and recycling of poisonous warfare substances will allow provision of the objective ecological risk analysis and regulation.

References

[1] Guidelines for hazard evaluation procedures (2-nd ed.).(1992). CCPS. American Institute of Chemical Engineers. New York.

[2] Chukhin, S.G. (1991). Social-Economical Criteria of Acceptable Radiation Risk of New Radiation Technologies. Energoatomizdat. Moscow.

[3] Kovalev, E.E. (1975). Basis of Concept of Acceptable Risk. Issues of Dosimetry and Radiation Protection 14:5.

[4] Elokhin, A.N. (1994) To the Issue of Acceptable Risk Criteria Definition. Problems of Safety during Emergency. 8:42-51.

[5] Katulski, Y.N. (2001). Disease Risk Assessment during System Toxicants Action. Hygiene and Sanitation 6:66-68.

[6] Rybolovlev, Y.R., Rybolovlev, R.S. (1979). Substances Dosing for Mammal according to the Constants of Biological Activity. DAN 247(6):1513-1516.

REVISED AIRBORNE EXPOSURE LIMITS FOR CHEMICAL WARFARE AGENTS

John A. Decker[1], Harvey W. Rogers[1]
[1]*Department of Health and Human Services, Centers for Disease Control and Prevention, Atlanta, Georgia, U.S.A. 30341*

Abstract: The Centers for Disease Control and Prevention (CDC) is revising airborne exposure limits (AELs) for tabun (GA), sarin (GB), VX, and sulfur mustard (H, HT, and HD) for demilitarization workers and the general public. New exposure criteria include short-term exposure limits (STELs) and immediately dangerous to life or health (IDLH) values. These new criteria augment revised long-term criteria, which include worker population limits (WPLs) and general population limits (GPLs). The criteria were revised using various risk assessment approaches, including reference concentration (RfC), relative potency, categorical regression (CatReg), carcinogenicity potency, and the CDC/National Institute for Occupational Safety and Health (NIOSH) IDLH methods. CDC believes the revised criteria meet the goal of protecting workers and the public at potential airborne concentration levels below those that would result in adverse health effects from acute exposures and further protect against risk for effects from long-term exposure.

Keywords: exposure limits, chemical warfare agents

1. INTRODUCTION

The chemical warfare agents tabun (GA or ethyl N,N-dimethyl-phosphoramidocyanidate, CAS 77-81-6); sarin (GB or O-isopropyl-methylphosphonofluoridate, CAS 107-44-8); VX (O-ethyl-S-(2-diisopropylaminoethyl)-methylphosphonothiolate, CAS 50782-69-9), and sulfur mustard (H, HT, HD or bis(2-chloroethyl) sulfide, CAS 505-60-2) are stored and are in the process of being destroyed by the U.S. Department of Defense (DoD). International treaty and Public Law 99-145 (50 U.S.C. 1521) mandate that all chemical warfare agents be destroyed. Public Law 91-121 and Public Law 91-441 (50 U.S.C 1512) mandate that the U.S. Department of Health and Human Services (DHHS) review DoD plans for disposing these munitions and make recommendations to protect public health. Within this context, DHHS/CDC recently revised the airborne exposure criteria for GA, GB, and VX and has proposed revisions for sulfur mustard. The original airborne exposure criteria were recommended by CDC

V.M. Kolodkin and W. Ruck (eds.),
Ecological Risks Associated with the Destruction of Chemical Weapons, 279–287.

in 1988 [1]. The WPLs, STELs, GPLs, and IDLHs are used to protect workers and the general population during routine chemical demilitarization activities. Separately, an array of acute exposure guideline levels (AEGLs), which are categorized according to the severity of health impact on the general population, has been finalized recently through the National Research Council. The development of AEGLs, which are applicable during emergency events and for emergency planning purposes, is discussed elsewhere [2].

2. NERVE AGENTS – GA, GB, VX

The nerve agents GA, GB, and VX are organophosphate esters that were designed specifically to cause incapacitation or death in military use. These agents are among the most potent of all chemical warfare agents and have extraordinarily high levels of acute toxicity. In vapor or aerosol form, the nerve agents can be inhaled or absorbed through the skin or the eyes. As a liquid, they can be absorbed through the skin, conjunctiva, and upper gastrointestinal tract. The agents' toxicity is related primarily to their ability to inhibit acetylcholinesterase, which is a critical enzyme needed for nerve function [3]. Health symptoms may include runny nose, tightness in the chest, dimness of vision and pinpointing of pupils (miosis), difficulty breathing, drooling and excessive sweating, nausea, vomiting, cramps, involuntary defecation and urination, twitching, staggering, headache, confusion, drowsiness, coma, and convulsions. The signs and symptoms of agent toxicity may be followed by cessation of breathing and death [4].

The EPA reference concentration (RfC) risk assessment method was used to derive the exposure criteria for GB [5]. The risk assessment attempts to extrapolate potential minimal or no observed biologic effects in humans at low-level exposures where such epidemiologic or toxicologic data are not directly available. This method for deriving exposure criteria has evolved over 30 years. The critical health effect used to derive the exposure criteria for GB is miosis, which occurs as a direct vapor effect on the eye before depression of red blood cell acetylcholinesterase [6].

Recommended AELs for GB: CDC recommends a new WPL value of 3×10^{-5} mg/m^3, expressed as an 8-hour time-weighted average (TWA). Additionally, CDC recommends a STEL of 1×10^{-4} mg/m^3 to be used in conjunction with the WPL. Exposures above the WPL up to the STEL should not be longer than 15 minutes and should not occur more than 4 times per day, and at least 60 minutes should elapse between successive exposures in this range. The STEL should not be exceeded during the work day, even if the cumulative exposure over the 8-hour TWA is not exceeded. CDC recommends a decrease in the GPL to 1×10^{-6} mg/m^3. The WPL and GPL values are approximately threefold lower than those values previously recommended by CDC in 1988 [1]. An IDLH value of 0.1 mg/m^3 is recommended for GB.

Relative potency generally is not regarded as an optimal method for deriving exposure criteria because the dose-response curves between the candidate chemical and the reference chemical may not be proportional; however, specific circumstances, such as lack of applicable toxicity data, may support using relative potency. Because the toxicity databases for GA and VX were inadequate for deriving exposure limits, these criteria were derived by evaluating their relative potency compared to that of GB.

Recommended AELs for GA: Although not as well-studied as GB, GA is thought to be approximately equal in potency to GB. Therefore, CDC recommends same exposure limits for GA as for GB.

Recommended AELs for VX: CDC recommends exposure limits for VX derived using a relative potency of 12 times greater than GB, with application of a modifying factor of 3 for the incomplete VX data set. CDC recommends that the VX WPL, expressed as an 8-hour TWA, should be decreased to 1×10^{-6} mg/m^3. Additionally, CDC recommends a VX STEL of 1×10^{-5} mg/m^3. An excursion to the VX STEL should not occur more than one time per day (compared with four times per day for a typical STEL). The recommended WPL is a factor of 10 lower than the CDC's 1988 recommendation [1]. CDC recommends that the GPL for VX be decreased to 6×10^{-7} mg/m^3 (a factor of five lower than CDC's 1988 recommendation). The VX GPL and STEL were adjusted in acknowledgment of the technical limitations of current air monitoring methods, while assuring that the GPL and STEL would protect public health. Further details can be found in the published proposal [6]. An IDLH value of 0.003 mg/m^3 is recommended for VX. CDC's recommended AELs, which are summarized in Table 1, are expected to be published in the *Federal Register* in late 2003.

Table 1: CDC Recommended Airborne Exposure Limits (AELs) for GA, GB, and VX in Demilitarization

AELs (mg/m^3)	General Population Limit (GPL)	Worker Population Limit (WPL)	Short-Term Exposure Limit (STEL) (Workers)	Immediately Dangerous to Life or Health (IDLH) (Workers)
GA, GB – Recommended	1×10^{-6}	3×10^{-5}	1×10^{-4}	0.1
GA, GB – Previous (1988)	*3×10^{-6}*	*1×10^{-4}*	--	*0.2 (Army)*
VX – Recommended	6×10^{-7}	1×10^{-6}	1×10^{-5}*	0.003
VX – Previous (1988)	*3×10^{-6}*	*1×10^{-5}*	--	*0.02 (Army)*
Averaging Time	24 hours	8 hours	15 minutes	≤30 minutes
Monitoring Method for Recommended Exposure Criteria	Historical monitor†	Historical monitor	Near-real-time monitor	Near-real-time monitor

* VX STEL was adjusted from 4×10^{-6} mg/m^3 (up to four times per day) as proposed in the *Federal Register* announcement to 1×10^{-5} mg/m^3 (not more than one time per day) based on technical capabilities of existing air-monitoring technologies.

† Historical monitoring typically refers to long-term sampling and analytical methods. Air-monitoring results from historical methods are not known until laboratory analyses are complete.

3. BLISTER AGENT - SULFUR MUSTARD (H, HT, AND HD)

Sulfur mustard's toxicity as a blister agent and irritant results primarily from its ability to form covalent linkages, attaching its alkyl group to various cell constituents. However, the exact means by which sulfur mustard causes chemical and physical damage is still not well understood. Humans respond to skin application by blistering, and the distinctive blisters common to mustard casualties seem to depend on some property unique to human skin [7]. Sulfur mustard affects any epithelial surface it contacts, with both vapor and liquid forms readily penetrating thin layers of most fabrics to reach underlying skin. Its lipophilicity ensures effective absorption into intact skin. Sulfur mustard's rapid skin penetration is enhanced by moisture, heat, and thin skin. Because of the rapid fixation of mustard to tissue, the fluid inside the blisters that eventually develop contains no free mustard, posing no secondary contamination risk to health-care providers.[8] Sulfur mustard can volatize, generating vapors with a characteristic garlic-like odor at typical room temperature. Exposure to vapor can irritate or damage the eyes, respiratory tract, and skin. Effects follow an asymptomatic latency period of hours or days. Testing on both laboratory animals and humans indicate that ocular effects are the critical toxic effect in that no overt systemic involvement is seen without effects on the eye. However, deaths from mustard exposure often result from secondary infections involving the respiratory tract. Sulfur mustard agent was designated as a Group 1 known human carcinogen by the International Agency for Research on Cancer over 20 years ago [9]. Current understanding of the mechanisms of carcinogenicity and recent reviews of HD carcinogenicity suggest that cancer is unlikely to develop after short-term, low-dose exposures to sulfur mustard [10,11,12].

The proposed exposure criteria for sulfur mustard were developed through a consideration of several risk assessment methods (RfC, CatReg, carcinogenicity potency, and the CDC/NIOSH IDLH method) within the context of applying professional judgment. CatReg is a recent risk assessment computer program that considers a broad array of exposure data and reported health effects to identify a "point of departure" on the basis of a family of concentration times time (Ct)-response curves[12,13]. The NIOSH IDLH risk assessment method is based on an approximation of the one-half hour lethal concentration for healthy workers divided by a factor of ten [14].

Proposed AELs for sulfur mustard (H, HT, and HD): CDC proposes that the 1988 WPL limit of 0.003 mg/m^3, currently an 8-hour TWA (WPL-8), becomes a 5-minute ceiling limit value (Ceiling-5M); and the GPL of 0.0001 mg/m^3, currently a 72-hour TWA (GPL-72), becomes a 12-hour TWA (GPL-12), and it is adjusted to 0.00002 mg/m^3 to meet carcinogenicity protection levels below a threshold of significant risk (Table 2). To ensure that undetected low-level exposure is not occurring, CDC also proposes implementation of historical monitoring at the new GPL-12 where workers are assigned, if a reasonable potential exists for sulfur mustard exposure. The proposed IDLH value of 0.7 mg/m^3 was derived in accordance with CDC/NIOSH protocol. CDC is receiving public comments on the proposed HD criteria.

Table 2: CDC Proposed Airborne Exposure Limits (AELs) for Sulfur Mustard*

AEL (mg/m^3)	**General Population Limit (GPL-12) (General Population and Workers)**	**5-Minute Ceiling Limit (Ceiling-5M) (Workers)**	**Immediately Dangerous to Life or Health (IDLH)† (Workers)**
HD - Proposed	0.00002	0.003	0.7
HD – Existing (1988)	*0.0001 (72 hours)*	*0.003 (8-hour TWA)*	-
Averaging Time	12 hours	≤5 minutes	<30 minutes
Monitoring Method	Historical, analysis as soon as practicable; not to exceed 72 hours after sample collection‡	Near-real-time monitor	Near-real-time monitor
Bases for Criteria	Exposure limit derived from consideration of carcinogenicity, RfC and CatReg.	Exposure limit derived from consideration of existing risk management practice, RfC, CatReg, and carcinogenicity (assuming continuous low-level background exposure for carcinogenicity)	Exposure limit derived by NIOSH according to protocol in *Documentation for Immediately Dangerous to Life or Health Concentrations.*

* Includes H, HD, and HT, collectively referred to as sulfur mustard
† Requires highly reliable dermal and respiratory protection providing maximum worker protection. The 30-minute period is not meant to imply that workers should stay in the work environment any longer than necessary, in fact, every effort should be made to exit immediately.
‡ Historical workplace monitoring should be conducted in potential HD-contaminated areas where workers are stationed.

4. CONCLUSIONS

The proposed airborne exposure criteria for chemical warfare agents are intended to protect workers and the general public from potential acute and long-term adverse health effects during chemical weapons disposal. The criteria are neither numerically precise values that differentiate between nonharmful and dangerous conditions, nor are they precise thresholds of potential human toxicity. Furthermore, the recommended changes to the criteria do not reflect change in, nor a refined understanding of, demonstrated human toxicity to these substances but rather the changes proposed resulted from updated risk assessment methodology and minimally modified risk assumptions.

References

[1] CDC. Final Recommendations for Protecting the Health and Safety against Potential adverse Effects of Long-Term Exposure to Low Doses of Agents: GA, GB, VX, Mustard Agent (H, HD, HT), and Lewisite (L). Federal Register, March 15, 1988; 53:8504-8507.

[2] National Research Council. Acute Exposure Guideline Levels for Selected Airborne Chemicals. Washington, DC: National Academies Press, 2003.

[3] Munro NB, Ambrose KR, Watson AP. Toxicity of the Organophosphate Chemical Warfare Agents GA, GB, and VX: Implications for Public Protection. Environ Health Perspect 1994;102:18-38.

[4] Department of the Army. Textbook of Military Medicine. Washington, DC: Office of the Surgeon General at TMM Publications, Bordon Institute, Walter Reed Army Medical Center, 1997.

[5] U.S. Environmental Protection Agency. Methods for Derivation of Inhalation Reference Concentrations and Application of Inhalation Dosimetry. Washington, DC: U.S. Environmental Protection Agency, October 1994; Report No. EPA/600/6-90/066F.

[6] CDC. Airborne Exposure Limits for Chemical Warfare Agents GA (tabun), GB (sarin), and VX. Federal Register January 8, 2002; 67:894-901.

[7] Marrs TC, Maynard RL, Maynard RL, Sidell FR. Chemical warfare agents: toxicology and treatment. New York: John Wiley and Sons, 1996.

[8] U.S. Army. Medical Management of Chemical Casualties Handbook. Aberdeen Proving Ground: Medical Research Institute of Chemical Defense (MRICD). 1999.

[9] International Agency for Research on Cancer. IARC Monographs on the Evaluation of Carcinogenic Risk of Chemicals to Man: Some Aziridines, N, S, & O-mustards and Selenium, vol. 9. Lyons, France: International Agency for Research on Cancer, 1975:181-207.

[10] U.S. Environmental Protection Agency. Guidelines for Carcinogen Risk Assessment. Review Draft, NCEA-F-0644, Washington, DC: Risk Assessment Forum, July 1999.

[11] Bullman T, Kang H. A fifty-year mortality follow-up study of veterans exposed to low-level chemical warfare agent, mustard gas. Ann Epidemiol 2000;10:333-338.

[12] Hurst CG, Smith WJ. Chronic effects of acute low-level exposure to the chemical warfare agent sulfur mustard. In: Somani SM, Romano JA, Eds. Chemical Warfare Agents: Toxicity at Low Levels. New York: CRC Press, 2001.
[13] U.S. Environmental Protection Agency. CatReg User Software Manual, EPA/600/R-98/052F, Washington, DC: Office of Research and Development, 2000.
[14] U.S. Environmental Protection Agency. CatReg Software Documentation, EPA/600/R-98/053F, Washington, DC: Office of Research and Development, February 2000.
[15] National Institute for Occupational Safety and Health. Documentation for Immediately Dangerous to Life or Health Concentrations. Cincinnati, Ohio: National Institute for Occupational Safety and Health, NTIS Publication No. PB-94-195047, 1994.

ANALYSIS OF TECHNOLOGIES FOR LEWISITE DESTRUCTION

Vadim Petrov[1], Aleksey Trubachev[1], Aleksey M. Lipanov [1]
[1]*Institute of Applied Mechanics UB RAS, Izhevsk (Russia)*

Abstract: In Kambarka (Udmurt Republic, Russia) there is the world's largest arsenal of arsenical poisoning substance – lewisite. The total amount of lewisite is about 6,400 tons. At the present time, from the all existing technologies for lewisite destruction only two have been chosen as the main ones. The first one is alkaline hydrolysis of lewisite, followed by the electrolysis of the reaction masses and high temperature ammonolysis of lewisite. The final product of both technologies is metallic arsenic of high purity for electronic engineering. However, the subject analysis of the given technologies for lewisite destruction shows a number of essential technical difficulties, which can affect the safety of processes under the condition of large scale performance. In this connection, it seems more reasonable to use less complicated and more efficient technologies, which do not yield metallic arsenic, but at the same time they do not form any highly toxic intermediate and final products. The technology that produces arsenic sulphide can be considered the more suitable one.

Key words: lewisite, destruction, safety

1. INTRODUCTION

We have already presented to some extent the assessment of the technologies for destruction of lewisite in Russia [1]. In this paper we give a more detailed analysis and express our opinion as specialists in this sphere and deal directly with this problem. It should be remembered that the world's largest storage of lewisite is situated in our district, the Kambarsky district in the Udmurt Republic.

2. MATERIALS AND DISCUSSION

When considering the technologies for destruction of lewisite we have used pre-project documentation of the facility for lewisite destruction in the city of Kambarka, the Udmurt Republic. These materials were presented to the members of the specialized commission, which is part of the

V.M. Kolodkin and W. Ruck (eds.),
Ecological Risks Associated with the Destruction of Chemical Weapons, 289–296.

Government of the Udmurt Republic [2, 3], for consideration. Also we have used various publications [4 - 6] and the results of our work on the development of the technologies for processing of reaction masses of lewisite detoxification [1, 7, 8].

As is known, according to the adopted program on chemical weapons destruction in Russia, the facilities for chemical weapon destruction will be implanted place in the regions where they are stockpiled, because transportation for large distances over the territory of Russia is forbidden. This fact can considerably increase the cost of the facilities for chemical weapon destruction, because most of the regions that stockpile chemical weapons, are industrially backward territories. Therefore at the initial stage it is necessary to inject money not directly into construction of these facilities for chemical weapon destruction, but into creation of the industrial infrastructure of these facilities. Also it is necessary to mention the lack of possibility of hiring specialised personnel for such facilities among local inhabitants. This description applies to the Kambarsky district in the Udmurt Republic. According to the official data [9] in Kambarsky district there are only nine industrial projects, and none of them has chemical production. The largest industrial project is the Kambarsky Machine-building plant, its main activity being metal working.

The approach to establishment of the technologies for destruction of lewisite, specifically, can be formulated in accordance with this. The technologies should be suitable for the existing regional social-economic conditions. Only in this case is it possible to guarantee the fulfillment of such a project with a high degree of probability. According to all of the preceding, there is a reason to believe that the preference should be given to simple and reliable technologies, which are more ecologically safe and have justified design.

As is known, the completed competition for choosing the technology of lewisite destruction [10] was reduced to the problem of lewisite decomposition and to producing metallic arsenic from it. Metallic arsenic was supposed to be used for a source of arsenic of high purity for electronics. It was thought that such an approach would allow for compensation for part of the expenses for the program of chemical weapon destruction in Russia. However, the study of the demand for metallic arsenic and arsenic of high purity in the domestic and external markets has shown a lack of necessity of processing lewisite into metallic arsenic. For instance, the demand for arsenic of high purity in the domestic market makes up not more than ten tons per year. Out of the lewisite in Kambarka it is possible to receive about 2500 tons of metallic arsenic. Thus, such an approach cannot be justified. To keep in store a large amount of arsenic for a long period of time is also unreasonable, as far as there is no deficiency of arsenic at all.

Therefore, the decision on the technologies for lewisite destruction cannot be considered as a final one. The technologies, that give some other final products and that are suitable for regional peculiarities, should be also thoroughly considered.

In this paper we present the comparison of three technologies of lewisite destruction and lewisite utilization and give the assessment of these technologies from the point of view of their fitness for the regional peculiarities. The first technology is lewisite destruction by alkaline hydrolysis followed by electrolysis of reaction masses and production of metallic arsenic [2]. This technology is being developed by the State Research Institute of Organic Chemistry and Technologies (Moscow). The second technology is high-temperature ammonolysis of lewisite with receiving of metallic arsenic [3], being developed by Nizhny Novgorod State University. The third technology is alkaline hydrolysis of lewisite followed by precipitation of arsenic sulphide from reaction masses [1], being developed by the Institute of the Applied Mechanics of the Ural Branch of Russian Academy of Sciences (Izhevsk). The first two technologies have been approved during the competition of the technologies [10]. The last one is an alternative to the first two.

3. BRIEF DESCRIPTION OF THE TECHNOLOGIES

3.1 Alkaline hydrolysis of lewisite followed by electrolysis of reaction masses and production of metallic arsenic (1)

The reaction of lewisite decomposition by alkali solution can be written in the form:

$$C_2H_2\,AsCl_3 + 6NaOH \rightarrow C_2H_2 + Na_3AsO_3 + 3NaCl + 3H_2O \qquad (1)$$

The reaction is carried out at the temperature 95 - 104 C for 0.5 hour [4,5]. Acetylene, formed during the reaction, is removed by the inert gas purge and then it is burnt. The produced reaction mass is subjected to electrolysis. The electrochemical process can be written in the following form:

$$4Na_3AsO_3 + 12NaCl + 17H_2O \rightarrow 2As + 2AsH_3 + 2H_2 + 24NaOH + 6Cl_2 + 2.5O_2 \qquad (2)$$

As a result of the process a pulp of metallic arsenic is produced; it contains sodium arsenite, sodium hydroxide, sodium chloride. The pulp is isolated from catholyte, washed and dried in vacuum. Metallic arsenic is placed into a glass container, hermetically sealed and stored. Remaining catholyte, enriched with sodium hydroxide, is directed to the lewisite alkaline hydrolysis stage. At the cathode such gaseous products as hydrogen and arsine are formed. Trapping of arsine is performed in a scrubber by solution of arsenic chloride. At the anode oxygen and chlorine are formed; chlorine is trapped by alkaline solution with production of a solution of hypochlorite.

3.2 Processing of lewisite into elemental arsenic by the method of high temperature ammonia reduction (2)

The considered process is presented as follows. On interaction of lewisite with ammonia in the reactor at the temperature 650 - 750 C, elemental arsenic, hydrocarbons and ammonium chloride are formed [6]. The reaction can be given in the form:

$$C_2H_2AsCl_3 + 4NH_3 \rightarrow As + 3NH_4Cl + C_2H_2 + 0.5N_2 \qquad (3)$$

The resulting products are transferred into the scrubber, where they are processed by water and then collected in a vessel. Volatile hydrocarbons are purified and burnt. Ammonium chloride is dissolved in water and separated from metallic arsenic. Metallic arsenic is washed, dried and stored. Solution of ammonium chloride is evaporated to dryness and marketable ammonium chloride is produced.

3.3 Alkaline hydrolysis of lewisite with precipitation of arsenic sulphide from reaction masses (3)

Alkaline hydrolysis of lewisite is performed according to reaction (1). Into the produced reaction mass the solution of sodium sulphide and hydrochloric acid is added, which results in quantitative transition of arsenic into the form of sulphide, and a solution of sodium chloride is formed [1,7,8]. The process proceeds under usual conditions with high rate, when the values of pH are within the interval from 2 to 4. The reaction can be written in the following form:

$$2Na_3AsO_3 + 3Na_2S + 12HCl \rightarrow As_2S_3 + 12NaCl + 6H_2O \qquad (4)$$

Arsenic sulphide is filtered, washed and dried. The produced arsenic sulphide is fused into insoluble blocks and transferred to the specialized plant for further processing. Solution of sodium chloride is evaporated down to a concentration of 300 g/l and directed to the plant for electrochemical production of sodium hydroxide and hydrochloric acid.

4. COMPARISON OF THE TECHNOLOGIES

Let us compare several parameters of the technologies:

Name of the parameter	Technology		
	1	2	3
1. Capacity, tons per 24 hours	5	5	10
2. The number of working days per year	300	300	150
3. The necessity to work during a cold season	yes	yes	no
4. The time for destruction of the whole amount of lewisite, amount of years	4.5	4.5	4.5
5. Limiting stage of the process	electrolys.	ammonolys.	alk. hydrolys.
6. Presence of toxic products	arsine clorine	ammonia	no
7. Presence of fire hazard products	acetylene hydrogen qxygen	acetylene volatile hydrocarbons	acetylene
8. Electric energy, kWh per 1 ton of lewisite	15700	589	< 300
9. Personnel	553	377	250
10. Personnel with high qualification, %	70	80	50
11. Capital expenditure, %	100	70	60
12. Final product	metallic As	metallic As	As sulphide

13. Special conditions for storing	necessary	necessary	not necessary
14. Standard equipment, %	40	20	80

As can be seen from this comparison the 3-rd technology has certain advantages compared to the other two. And first of all it addresses the problem of security.

On the electrolysis of reaction masses according to the first technology 35.5 kg of fire hazard and explosive arsine - hydrogen mixture per 1 ton of lewisite are formed, containing 45%mass of hydrogen and 54%mass of arsine. Besides, arsine has strong toxic properties. Its maximum allowable concentration in the atmosphere of the working zone is 0.1 mg/m^3. Despite the fact, that arsine is supposed to be trapped, there is still potential danger for the personnel.

The same can be said about the second technology that deals with ammonia: it demands special precautionary measures. It means that both technology 1 and technology 2 involve additional risks, including the risk connected with the work with lewisite itself as toxic substances are used or are formed.

On the other hand, such processes as electrolysis and ammonolysis of lewisite have not been studied enough, and have been used mainly under laboratory conditions. The technology of arsenic sulphide precipitation, analogous to the one we have offered for detoxification of lewisite reaction masses, has been widely used at the mining-metallurgy plant (Karaganda, Republic of Kazakhstan). This fact presents the opportunity to considerably reduce the amount of time and finance for carrying out research and experimental works for testing the third technology, and also for mounting the installation, as far as this installation includes mainly standard equipment. According to our assessment, it takes 1.5 - 2 years for research efforts together with mounting the installation.

The final product of the third technology is arsenic sulphide, which is not going to be accumulated or stored in Kambarka; instead it will be directed to Chelyabinsk region for further processing. The produced brine of sodium chloride with concentration 300 g/l and higher is planned to be directed to electrolysis plant in Perm region. All these plants are not farther than 300 - 400 km from Kambarka. Thus in Kambarka it is necessary to build only specialized plant for lewisite destruction and for the initial process stage of the reaction masses. This will allow for the considerable reduction of the capital expenditures for the construction of the facility and increase its efficiency.

The other technologies propose storing metallic arsenic in Kambarka. In this case there is a risk of penetration of arsenic and its compounds into the environment and also there is a certain risk connected with the occurrence of fire hazard and explosive situations.

5. CONCLUSION

In this regard, the chemical weapons destruction the preference should be given to simple, reliable and thoroughly developed technologies, which has high efficiency, and do not form toxic and explosive intermediate and final substances, and do not require large capital expenditures. As can be seen from the above-mentioned comparison the technologies for producing metallic arsenic do not meet such requirements.

References

[1] Petrov, V.G., Trubachyev, A.V., Khan, V.P. (1998). Processing of solutions, formed at destruction lewisite, with production of arsenic sulphide. In: Proceedings of the 6-th CBW Protection Symposium, Stockholm, Sweden, pp. 341-343.

[2] Technico - economic justification. Industrial zone of the plant for lewisite destruction in Kambarka, the Udmurt Republic. Variant 1. (1995). Institute for Projecting Plants of Organic Synthesis "Giprosintez", Volgograd.

[3] Technico - economic justification. Industrial zone of the plant for lewisite destruction in Kambarka, the Udmurt Republic. Variant 2. (1995). Institute for Projecting Plants of Organic Synthesis "Giprosintez", Volgograd.

[4] Umyarov, I.A., Kuznetsov, B.A., Krotovich, I.N. (1993). The methods of destruction and utilization lewisite and yeprite. Russian chemical journal 38:3:25-29.

[5] Petrunin, V.A., Baranov, Yu.I., Kuznetsov, B.A. (1995). Mathematical modeling of process for alkali hydrolysis of lewisite. Russian chemical journal 39:4:15-17.

[6] Zorin, A.D., Karatayev, Ye.N., Klimov, K.N. (1994). Gas processing of lewisite in metallic arsenic and arsenic oxide. Vestnik of the Udmurt University, Izhevsk, Russia, pp. 22-41.

[7] Petrov, V.G., Trubachev, A.V., Khan, V.P. (1995) Method for lewisite destruction. Patent of the Russian Federation № 2099116

[8] Petrov, V.G., Khan, V.P., Trubachev, A.V., Chechina, A.A. (2003) Decisions adopted in connection with the storage of arsenic-containing products formed as the result of lewisite destruction. In: Proceedings of the NBC 2003 International Symposium, Jyvaskyla, Finland, pp. 197-210.

[9] Reshetnikov, S.M., Makarova, L.L., Kornyev, B.I. (1994) Results of research soils and natural water reservoirs in 30 kilometer zone of Kambarka city on the contents in them of arsenic and some heavy metals. Vestnik of the Udmurt University, Izhevsk, Russia, pp.109-121.

[10] Petrov, S.V. (1995). Expert estimation of technologies for destruction of lewisite stocks. Russian chemical journal. 39:4:4.

NEW UNDERSTANDING ON PATHOGENESIS OF DELAYED EFFECTS OF RVX LOW-DOSE CHRONIC EXPOSURE

Nikolay Goncharov[1], Andrey Radilov[1], Igor Mindukshev[2], Sergey Kuznetsov[2], Yelena Yermolayeva[1], Lidia Glashkina[1], Irina Shkayeva[1], Irina Dobrylko[2], Anatoly Kuznetsov[1]

[1]*Institute of Hygiene, Ocupational Pathology and Human Ecology, St.Petersburg (Russia),*
[2]*I.M.Sechenov Institute of Evolutionary Physiology and Biochemistry. St.Petersburg (Russia)*

Abstract: In the context of carrying out CW non-proliferation conventional programs, chronic effects of subsymptomatic concentrations of highly toxic OP available in the stockpiles of chemical agents are of particular interest. During 3 months rats of the 1-5 groups consumed daily RVX with drinking water in a dose from 10^{-6}g to 10^{-10}g per 1 kg body weight, respectively. The comparative analysis of biochemical and physiological parameters studied is indicative of the complete absence of significant changes of the activity of plasma and erythrocyte cholinesterases in rats of all groups. The functional activity of platelets was investigated by a novel method of low angle light scattering. Kinetic parameters of aggregation are changed after 3 months of chronic intoxication, 2 and 6 months after stopping of intoxication with RVX. The latent period of M-response, the rate of rising action potential, and its duration, all change significantly. Non-specific reaction of leukocytes was reliable in groups 1-3. The results presented allow us to assert the key role of non-synaptic mechanisms in developing effects of RVX chronic action. The functional changes at the microcirculatory level may prove to be the most significant factors in the etiology of delayed effects.

Key words: RVX, organophosphates, platelets, nerves, immunity.

1. INTRODUCTION

At present three types of damages induced by organophosphates (OP) are distinguished: acute poisoning, the so-called intermediate syndrome, and OP-induced delayed polyneuropathy (OPIDP) [1]. The acute poisoning mechanisms are realizable via acetylcholinesterase (AChE) inhibition. The intermediate syndrome is considered to be a delayed complication of severe acute poisoning, and OPIDP may be the consequence of both acute and chronic poisoning with some OP - neuropathic agents, the typical

V.M. Kolodkin and W. Ruck (eds.),
Ecological Risks Associated with the Destruction of Chemical Weapons, 297–303.

biochemical sign of their action being inhibition of the so-called neurotoxic esterase (NTE).

In the context of carrying out CW non-proliferation conventional programs, chronic effects of subsymptomatic concentrations of highly toxic OP prevalent in the stockpiles of chemical agents are of particular interest, since the risk of contact of personnel of CW destruction facilities with chemical agents cannot be ruled out. The problem consists of a search for new criteria of intoxication. One should differentiate at least two aspects of this problem - molecular and functional. The former consists of the appearance of new molecular targets of OP action, other than AChE, butyrylcholinesterase (BChE) or NTE. For instance, carboxylesterases (CarE) may be rather sensitive targets compared to AChE in cases of chronic action of OP [2-6]. The second (functional) aspect of this problem concerns the development of pathologic symptoms, the molecular causes of which are not due to synaptic AChE inhibition. One of the first cases of OP non-cholinesterase effects (when the interrelation between the OP molecular target and the functional disturbance was observed) involved inhibition of kinurenin formamidase of the hen egg yolk sac membrane responsible for teratogenic effects [7]. Another example of OP teratogenic effect is abnormal development of the conjunctive tissue of Xenopus embryos due to inhibition of lysiloxidase and incomplete post-translational modification of collagen [8]. Of particular interest is the evidence of AChE activity over the region of the basal membrane of capillary endothelium. Additionally, the cholynacetyltransferase activity has been found on the luminal surface of endothelial cells [9]. The presence of muscarinic receptors on endothelium supports the role of endothelium as a peculiar cholinergic structure. Moreover, endothelium was shown to transform parathion to paraoxon with cytochrome P450 [10]. After platelets were discovered as biochemical markers of delayed neurotoxicity, because of NTE found in these cells correlating with NTE activity in nervous tissue [11, 12], a question arose: Are there functional consequences of NTE inhibition in platelets, or is it just a “side-effect” of OP intoxication? Previously assumptions were offered about the direct influence of OP on the humoral components of hemostasis and the complement system because of the abundance of serine proteases in these systems [13, 1]. However, there are no data on functional consequences of OP low dose chronic action on these systems. There is no evidence on the interrelation of hemostasis and neuromuscular conduction either, which is of undoubted interest considering the low “percentage” of endoneural capillaries in the structure of nerve fibers and the dependence of the trophism of any tissue upon the integrity of the endothelial monolayer and adequate functioning of hemostasis. Finally, to understand mechanisms of OPIDP development more clearly, the comparative analysis of molecular and cellular mechanisms of pathogenesis of other neuropathies is necessary. This will help to understand whether OPIDP is an exclusively specific phenomenon or this disease has common features with other types of neuropathies. This work is one of the first attempts to summarize existing questions with our own experimental data. We have tried to model the chronic intoxication with RVX to find a correlation of some biochemical, physiological and immunological data in a long-term experiment on rats.

2. METHODS AND RESULTS

O-ethyl-S-(2-diisopropylaminoethyl)-methylthiophosphonate (RVX) was daily dissolved in drinking water to the concentrations of $5x10^{-6}$ (#1), $5x10^{-7}$ (#2), $5x10^{-8}$ (#3), $5x10^{-9}$ (#4), $5x10^{-10}$ (#5) g/100ml. The 5 groups of rats consumed daily 20 ml of RVX aqueous solution. So, during 3 months test animals of the 1st group consumed daily RVX with drinking water in a dose of 10^{-6}g, animals of the 2nd group - 10^{-7}g, of the 3rd – 10^{-8}g, of the 4th – 10^{-9}g, and those of the 5th group – 10^{-10}g per 1 kg body weight. Measuring cholinesterase activity in plasma and erythrocytes was performed by Ellman's method [14]. The functional activity of platelets was investigated by a novel method of low angle light scattering that allows all stages of the platelet transformation to be assessed [15]. Immunological status was estimated after a leukocytes' specific agglomeration test and a leukocytes' non-specific or spontaneous agglomeration test. Investigation of monosynaptic miotatic reflex and conduction rate through the peripheral nerve fibre was undertaken with *n.tibialis*.

The comparative analysis of biochemical and physiological parameters studied is indicative of the complete absence of significant changes of the activity of plasma and erythrocyte cholinesterases in rats of all groups relative to the control after exposure to RVX doses given above. However, after 3 months of intoxication, kinetic parameters of aggregation are significantly increased as to both normalized maximal rate *Umax* (in test groups ##1-3) and effective concentration EC_{50} (in the 1st and 2nd groups). Two months after stopping the chronic intoxication with RVX (rehabilitation period) reliable differences in both kinetic parameters were observed just in the group #1. Group #2 showed reliable increase of EC_{50} and just a tendency ($p<0.1$) to increase *Umax*. Group #3 showed a tendency ($p<0.1$) to increase EC_{50}. The final estimation of kinetic parameters of aggregation made after 6 months of rehabilitation showed reliable decrease of *Umax* in groups #1 and #2 and increase of EC_{50} in groups

##1, 3, and 4 (the latter showed no reliable differences at the earlier terms).

Study of immune reactions in the chronic experiment revealed weak specific sensitivity in the first group only. However, the non-specific reaction of leukocytes was reliable in groups ##1-3. These differences but less remarkable remained after 2 months of rehabilitation period. After 6 months of rehabilitation, only the tendency in the non-specific reaction of leukocytes was discovered in the three groups of animals. In groups ##4 and 5, no reliable changes were seen.

Because of stimulation of the peripheral nerve trunk of intact animals, muscle action potentials (AP) of three types occur: M-response (the result of the direct stimulation of α-motorneuron axons), H-response (the monosynaptic response proper), and polysynaptic responses with the variable latent period from 8-12 up to 40 msec and over. In test animals of the 1st group changes of temporal parameters refer chiefly to reduction of the latent period and duration of M-response. Polysynaptic responses occur at all intensities of excitation and have a more pronounced character than in intact rats. A tendency towards a slight increase of the conduction velocity through fibres is noted. This is most pronounced in fibres of Aδ group. In addition, the marked level and more distinct differentiation of the peaks of the composite action potential are noted. The results obtained in the 2nd group differ more significantly from the control. Along with normal AP (of "spike" type) there are slow waves of depolarization of up to 30 ms duration. Another significant difference is the absence of the "sub-threshold border" (graduation of the increase of the amplitude of AP when the irritation stimulus enhances). The latent period of M-response, the rate of rising AP, and its duration, all increase significantly. The results of testing rats of the 3rd group, including the latent period and AP duration, do not differ from the control. At the same time, it should be noted that some animals exhibit the entire spectrum of pathologic reactions observed in rats of the 1st and 2nd groups: fasciculation, the presence of slow (local, depolarized) potentials, and paradoxical discharges.

3. DISCUSSION

The results presented allow us to assert the key role of non-synaptic mechanisms in developing effects of RVX chronic action. The functional changes at the microcirculatory level may prove to be the most significant factor in the etiology of delayed effects. The changes observed are likely to reflect multiple microangiopathies in the compensatory phase. Long-term observations made previously by our physicians of victims of acute and chronic intoxications with RVX discovered vegetative changes in the cardiovascular system and "microorganic disorders" in the CNS of unknown etiology that could be of microvascular origin. Specific immune reactions, gastrointestinal disorders and hepato-biliary disfunctions (gastric mucous layer erosions and ulcers, toxic hepatitis), were also found. These observations, along with the results of our investigations and the analysis of literature data make it possible to suggest that in chronic oral administration of RVX in low doses events develop as follows. The non-specific covalent interaction of RVX with protein molecules of the plasma membrane and cytoskeleton causes primary functional changes. In addition, the RVX molecule acts as hapten and/or adjuvant and stimulates the immune reaction. It is possible that endotoxins are formed because of influence on intestine enterobacteria by way of developing anaphylactic and inflammatory reactions in the mucous layer. The most pronounced reactions should be expected at the first stages in microvessels of mucous membranes of the stomach and intestine (the "first barrier" targets), and in liver as well. At the following stages the pathogenesis is probably determined by two complementary factors: the immune reaction and tissue/circulatory hypoxia. The pathologic changes should be expected in tissues with slightly pronounced or no lymphatic drainage, as well as with high metabolic rates and oxygen demands: these are nerve fibres, CNS, and large blood vessels with *vasa vasorum* as their trophic vessels.

4. CONCLUSIONS

Kinetic parameters of platelet aggregation measured with low angle light scattering method are very sensitive functional criteria of chronic OP intoxication. The functional changes at the microcirculatory bed may prove to be the most significant factors in the etiology of delayed effects. The pathogenesis is probably determined by two complementary factors: the immune reaction and tissue/circulatory hypoxia. New understanding of the pathogenesis of delayed effects of OP intoxication may help to find a new medical treatment of these pathologies.

References

[1] Ray, D.E. (1998) Organophosphorus esters: An evaluation of chronic neurotoxic effects. De Montfort Univeresity Press, Leicester.
[2] Jewell, W.T., Miller, M.G. (1998) Identification of a caarboxylesterase as a major protein bound by molinate. Toxicol. Appl. Pharmacol. 149: 226-234.
[3] Krishnasamy, S., Gross, N.J., Teng, A.L., Schultz, R.M., and Dhand, R. (1997) Lung "surfactant convertase" is a member of the carboxylesterase family. Biochem. Biophys. Res. Commun. 235: 180-184.
[4] Poulsen, E., Aldridge, W.N. (1964) Studies on esterases in the chicken central nervous system. Biochem. J. 90: 182-189.
[5] Chemnitius, J.M., Zech, R. (1983) Inhibition of brain carboxylesterases by neurotoxic and nonneurotoxic organophosphorus compounds. Mol. Pharmacol. 23: 717-723.
[6] Richards, P.G., Johnson, M.K., Ray, D.E. (2000) Identification of acylpeptyde hydrolase as a sensitive site for reaction with organophosphorus compounds and a potential target for cognitive enhancing drugs. Mol. Pharmacol. 58: 577-583.
[7] Seifert, J., Casida, J.E. (1978) Relation of yolk sac membrane kynurenine formamidase inhibition to certain teratogenic effects of organophosphorus insecticides and of carbaryl and eserine in chicken embryos. Biochem. Pharmacol. 27: 2611-2615.
[8] Snawder, J.E., Chambers, J.E. (1993) Osteolathyrogenic effects of malathion in Xenopus embryos. Toxicol. Appl. Pharmacol. 121: 210-216.
[9] Lan, C.T., Shieh, J.Y., Wen, C.Y., Tan, C.K., Ling E.A. (1996) Ultrastructural localization of acetylcholinesterase and choline acetyltransferase in oligodendrocytes, glioblasts and vascular endothelial cells in the external cuneate nucleus of the gerbil. Anat. Embryol. (Berl). 194: 177-185.
[10] Delaunois, A., Gustin, P., Ansay, M. (1992) Altered capillary filtration coefficient in parathion- and paraoxon-induced edema in isolated and perfused rabbit lungs. Toxicol Appl Pharmacol. 116: 161-169.
[11] Husain, K., Mirza, M.A., Matin, M.A. (1991) Neurotoxic esterase activity in brain, spinal cord and platelets of certain birds and mammals. J. Appl. Toxicol. 11: 61-68.
[12] Husain, K. (1994) Phenyl valerate and cholinester hydrolases in platelets of human, hen, rat and mouse. Hum. Exp. Toxicol. 13: 157-163.

[13] O'Neill, J.J. (1981) Non-cholinesterase effects of anticholinesterases. Fundam. Appl. Toxicol. 1: 154-160.

[14] Ellman, G.L., Courtney, D.K., Andres, V., Jr., Featherstone, R.M. (1961) A new and rapid colorimetric determination of acetylcholinesterase activity. Biochem. Pharmacol. 7: 88-95.

[15] Деркачев, Э.Ф., Миндукшев, И.В., Кривченко, А.И., Крашенинников, А.А. (1998) Способ исследования активации и агрегации тромбоцитов. Патент RU 2108579 C1 6 G01 N 33/49. Б.И. №10 (II): 298.

HEALTH STATE AND HORMONE STATUS AMONG RESIDENTS OF KIZNER RESPECTIVELY WORKING AND NOT-WORKING AT ARSENAL OF CHEMICAL WEAPONS

Alexander Churakov[1], Nicolay Zabrodine[2], Ioulia Bouchmakina[2], Sverre Langard[3]

[1]The Izhevsk State Medical Academy, Izhevsk, (Russia), [2]The Centre for State Sanitary and Epidemiological Inspection in Udmurt Republic, Izhevsk (Russia), [3]The Centre for Occupational and Environmental Medicine, The National Hospital, Oslo (Norway)

Abstract: The relation between the possible influence of the arsenal of chemical weapons and health state and hormone status (thyrotropic, thyroxin, triiodtyronin, adrenocorticotrophic, hydrocortisone) among residents of Kizner both working and not-working at the arsenal of chemical weapons was established. Thus, the trend for emotional stress was determined for the people working at the arsenal of chemical weapons. The level of anxiety was higher for people working at the arsenal (40.67±0.96) in comparison with the people who did not work at the arsenal (39.71±1.00), mostly for men.

Key words: health state, hormone status, arsenal of chemical weapons

1. INTRODUCTION

The study was intended to establish the possible influence of the arsenal of chemical weapons on the health state and hormone status of the residents of Kizner. The emotional state, hormone status, and health state were evaluated for the residents of Kizner, both working and not-working at the arsenal of chemical weapons.

2. METHODS

The hormone status as an indicator of the emotional stress was evaluated at the week-start and at the week-end for people who worked (n=17) and did not work (n=16) at the arsenal of chemical weapons. The level of thyroid hormones (thyrotropic, thyroxin, triiodothyronin), adrenocorticotrophic, and

V.M. Kolodkin and W. Ruck (eds.),
Ecological Risks Associated with the Destruction of Chemical Weapons, 305–306.

hydrocortisone was measured using the radioimmunological analysis technique. In addition, the test for determination of the state of health, activity and mood was also applied. The level of anxiety was evaluated by the Spilberger test.

The collected data were evaluated by means of the Student K criterion and dispersion analysis.

3. RESULTS

The level of thyroid hormones and hydrocortisone was increased at the week-end ($P<0.05$) for all. The level of adrenocorticotrophic hormone was higher for the people working at the arsenal of chemical weapons (38.5±2.77 picogr/ml) in comparison with those not working at the arsenal (25.94±1.61 picogr/ml) ($P<0.05$). These results provide evidence of the possible high influence of the arsenal of chemical weapons on the people working at the arsenal. Functional hyperthyrosis was observed among the workers of the arsenal both at the week beginning and at the week-end. Those who did not work at the arsenal of chemical weapons had functional hyperthyrosis only at the week-end. Thus, the trend for emotional stress was determined for the people working at the arsenal of chemical weapons.

The level of anxiety was higher for the people working at the arsenal (40.67±0.96) in comparison with the people who did not work at the arsenal (39.71±1.00), mostly for men.

4. CONCLUSION

The relation between the influence of the arsenal of chemical weapons and the hormone status and health state among the residents of Kizner, both working and not working at the arsenal of chemical weapons, that had been only a hypothesis before, was established.

EVALUATION OF EMBRYOTOXIC AND TERATOGENIC EFFECTS OF FINAL PRODUCTS OF ORGANOPHOSPHORUS CHEMICAL WARFARE AGENTS (CWA) DESTRUCTION

Elena Ermolayeva[1]
[1]*Research Institute of Hygiene, Occupational Pathology and Human Ecology, St. Petersburg (Russia)*

Abstract: The destruction of chemical weapons by the two-stage technology requires accommodating great amounts of bitumen-salt masses (BSM) containing GB, GD, VX detoxication products in special-purpose storehouses at the landfill. During BSM long-term storage the probability of contamination of the surrounding environment with CWA destruction products due to leaching with rainfall and ground water must not be completely excluded. The embryotoxic activity (including teratogenic effects) of BSM aqueous extracts containing final products of CWA destruction was evaluated in white non-pedigree rats at concentrations 100%, 20%, and 4% of GB, GD, and VX. The main pathological changes were largely hemorrhages in internal organs. The toxicometry of BSM aqueous extracts allows 100% extracts to be classified as effective, 20% aqueous extracts – as possessing the threshold impact, and 4% aqueous extracts as non-effective for the majority of indices representing the state of the reproductive function. The most effective potential hazard has been observed for BSM-VX.

Key words: chemical weapon destruction, bitumen-salt masses, embryotoxicity, organophosphates.

Destruction of chemical weapons by the two-stage technology (detoxication and bituminization of neutralization products aimed at producing non-soluble bitumen-salt mass) requires accommodating great amounts of bitumen-salt masses (BSM) containing GB, GD, VX detoxication products in special-purpose storehouses at the landfill.

During BSM long-term storage the probability of contamination of the surrounding environment with CWA destruction products due to leaching with rainfall and ground water must not be completely excluded. It should be noted that the child's body is the most sensitive to the ecotoxicant effects, especially in the perinatal period.

The embryotoxic activity (including teratogenic effects) of BSM aqueous extracts containing final products of CWA destruction was evaluated in white non-pedigree rats at concentrations of 100%, 20%, and 4% of GB,

V.M. Kolodkin and W. Ruck (eds.),
Ecological Risks Associated with the Destruction of Chemical Weapons, 307–314.

GD, and VX, which in converting to the animal body weight correlates with 20, 4, and 0.8 ml/kg. The controls used were:

- drinking tap water;
- model water (simulating by its mineral content the ground water at the site of the supposed BSM storage (burial)), and
- bitumen aqueous extracts.

Bitumen-salt masses after GB and GD destruction include the following main components:

- calcium salts of isopropylmethylphosphonate,
- methylphosphonic acid,
- pinacolylmethylphosphonate,
- O-isopropyl - O' – (2 aminoethyl)methylphosphonate,
- O-pinacolyl – O' – (2 aminoethyl)methylphosphonate,
- monoethanolamine,
- bitumen,
- tributylamine,
- calcium fluoride, and
- calcium hydroxide.

After VX destruction BSM-VX consists of the following main components:

- potassium salts of isobutylmethylphosphonate and methylphosphonic acid,
- diisobutyl ester of methylphosphonic acid,
- N-methyl-pyrrolidone,
- έ - polymer of caprolactam,
- potassium diethylaminoethylmercaptide,
- potassium isobutylate, and
- bitumen [1].

The embryotoxic activity of BSM aqueous extracts was studied in accordance with methodical recommendations, currently in force in Russia (1986). The state of internal organs of fetuses was studied with the Wilson's microanatomic method [Wilson, 1973] modified by professor A.P. Dyban [Dyban et al., 1970]. The skeleton ossification was evaluated in embryonic total specimens stained by the Dowson method modified by Kimmel and Trammel (separate staining of cartilages and bones) [Kimmel et al., 1981]. A total of 298 litters have been examined and 2603 fetuses have been appraised.

One half of the fetuses from each litter was examined for the presence of gross pathological changes in internal organs, the remaining fetuses of each litter, for the presence of structural abnormalities of the skeleton development and rates of skeletal ossification.

BSM aqueous extracts were i/g administered for 1-22 days of pregnancy.

When evaluating embryotoxic action of BSM aqueous extracts the following results were obtained:

- the number of yellow bodies (corpora lutea) and the fertilization index in all test groups did not differ significantly from those in the controls;
- BSM aqueous extracts caused, in one degree or another, the decrease of the placenta weight and growth and weight parameters significant in case of exposure to 100% BSM-VX and BSM-GB extracts;
- the main embryolethal effect was the death of embryos in the preimplantation period; significant changes were seen in groups of animals that received 100% BSM-VX and BSM-GB aqueous extracts (Fig.1).

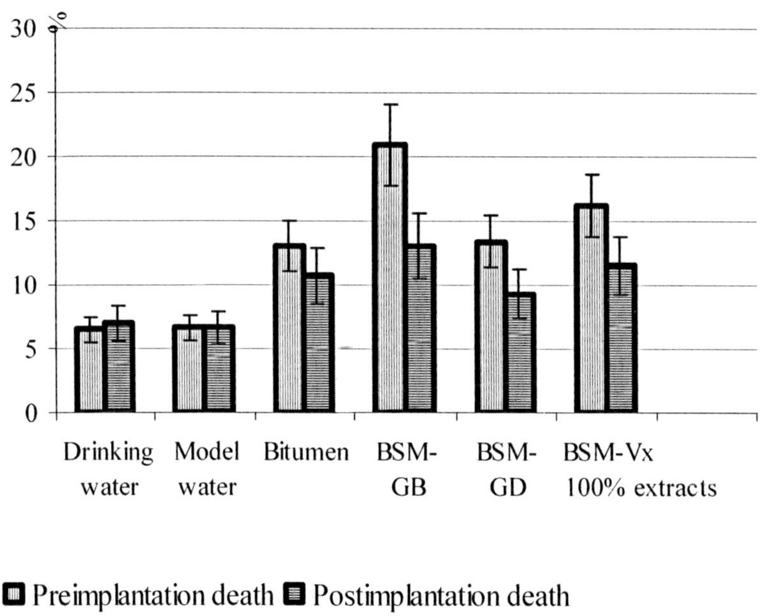

Fig.1. Pre- and postimplantation death of white rat embryos after intragastric administration of 100% BSM aqueous extracts for 1-22 days of pregnancy.

Visual examination of fetuses in control and test groups after exposure to BSM aqueous extracts revealed only single cases of gross pathological changes. The most abundant of them were small hematomas of different localization, hemostases, and hemorrhages in the umbilical cord and placenta junction. Among other pathological abnormalities single cases of edematous changes of the body, hemorrhagic exudate in fetal membranes, and anemia of tissue extremities were found. Cases of severe structural abnormalities – umbilical and cerebral hernias and cleft palate – were rather rare and were not of systemic nature.

The analysis of the state of internal organs showed that after BSM aqueous extracts exposure the number of litters containing fetuses with particular pathological changes increased with dose (Fig. 2).

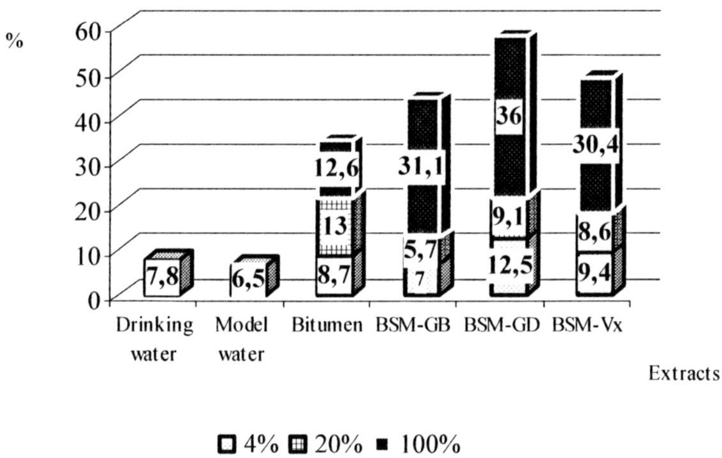

Fig. 2. The number of fetuses with pathological changes after exposure to 4. 20 and 100% concentrations of BSM aqueous extracts.

The main pathological changes were largely hemorrhages in internal organs. The analysis of hemorrhages localization sites demonstrated impairments of the head tissues, pleural cavity, pericardium, and abdominal cavity tissues (Fig. 3). Hemorrhages in kidney, liver, paravertebral spaces, and neck tissues were encountered much more rarely.

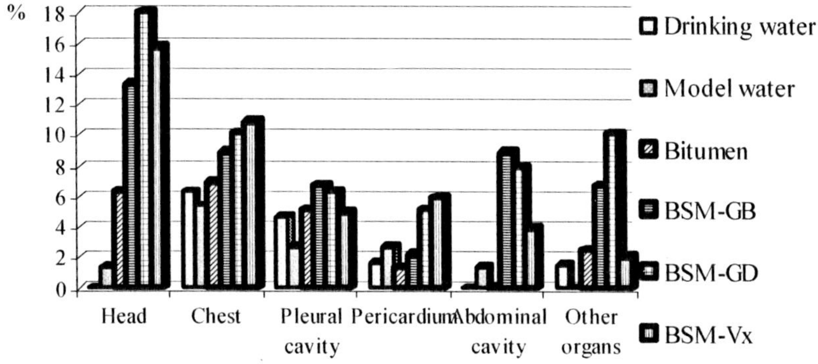

Fig. 3. Localization of hemorrhages in fetuses after exposure to 100% bitumen and BSM aqueous extracts.

The examination of the musculoskeletal system showed the absence of major structural abnormalities in fetuses of all test groups. The comparative analysis of the number of deformities among litters exposed to low, median and high concentrations of bitumen and BSM aqueous extracts did not exhibit any significant regularity between groups. The number of litters in which fetuses with skeletal pathology were noted comprised 10-15% of the total population. Significant effects were determined only after exposure to 100% BSM-VX, essentially as defects of tegumental bones of the skull without any systemic nature (Fig. 4).

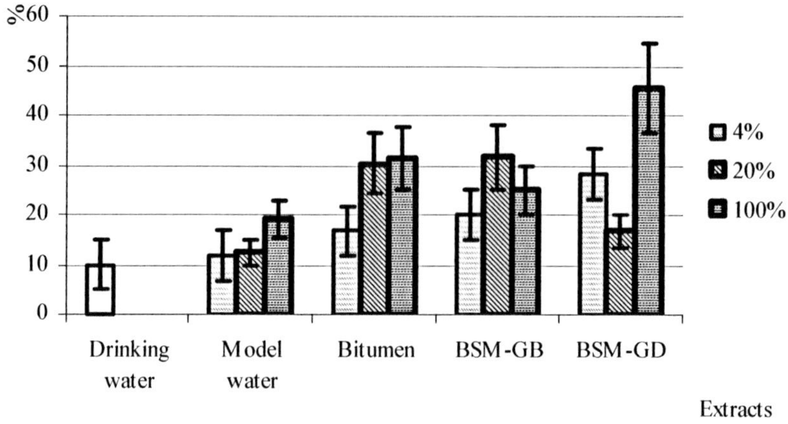

Fig. 4. Number of litters with pathological changes of skeleton in fetuses after exposure to BSM aqueous extracts.

The skeletal ossification is considered as a highly sensitive indicator that allows the normal rates of the fetal development to be compared with those after damaging exposure. The rat skeleton consists of 265-285 bones, but in the antenatal period deposition of calcium salts is not observed in all germs of bones. However, to appraise this process adequately, it is enough to evaluate the representative centers of ossification.

The majority of fetuses of all groups examined demonstrated sufficient synchronism and symmetry of the process of calcification of cartilaginous tissues about the body axis. The number of calcification centers in metacarpus and metatarsus elements was evaluated separately in right and left extremities. A significant increase of the number of ossification centers was noted in the form of replacement of cartilaginous tissue with osseous tissue in metacarpus and metatarsus elements, in the case of exposure to 20% BSM-GB and BSM-GD aqueous extracts. But more concentrated BSM aqueous extracts inhibited forming new ossification centers. So, evaluation of the ossification process of the skeletons of 2-day fetuses of the rat elucidated definite deviations from the control in forming rates of ossification centers in elements of fore and hind extremities (metacarpus, metatarsus) after exposure. The variations observed in the number of forming ossification centers are indicative of the weak response of ossification rates to BSM aqueous extracts exposure. These differences are not determinant for the subsequent formation of these skeletal elements and of the fetus as a whole; they define only some inhibition of ossification dynamics that can be easily compensated during the postnatal period of development.

Bitumen and BSM aqueous extracts possess weak or moderate embryotoxic properties. Pathological effects produced therewith are most often of a dynamic character and can be compensated in the course of the subsequent development.

Thus, in standard embryotoxic experiments on white non-pedigree rats it was determined that aqueous extracts of bitumen-salt masses containing GB, GD, and VX destruction products possess teratogenic activity largely expressed as hemorrhages. The time and dose dependent toxic effects in water-soluble unidentified components have been established. The toxicometry of BSM aqueous extracts allows 100% extracts to be classified as effective, 20% aqueous extracts as possessing the threshold impact, and 4% aqueous extracts as non-effective for the majority of indices representing the state of the reproductive function. The most effective potential hazard has been observed for BSM-VX, which requires more thorough study of biological effects produced by the destruction products of organophosphorus chemical warfare agents. At present we are studying the influence of organophosphorus chemical warfare agents on hemostasis. [Goncharov N.V. et al, 2003, Ermolaeyeva E.E. et al., 2003].

References:

[1] Ermolaeyeva, E.E. (2000) Evaluation of Delayed Effects of Final Products of Organophosphorus Chemical Warfare Agents Destruction. Author's abstract of the thesis. Saint-Petersburg, pp 23.

[2] Methodical Instructions for Studying Embryotoxic Action of Drugs and their Effects on the Reproductive Function. (1986). Pharmacological Committee attached to Public Health Ministry of the USSR, M., p. 24.

[3] Dyban, A.P., Baranov, V.S., Akimova, I.M. (1970) Main Methodical Approaches to Testing Teratogenic Activity of Chemical Compounds.- Archives of Anatomy, Histology and Embryology. №10, pp 89-100.

[4] Kimmel, C.A.,Trammel, C.A. (1981) A rapid Procedure for Routine Double Staining of Cartilage and Bone in Fetal and Adult Animals. Stain Technol., № 58, pp 271-273.

[5] Wilson, J.G. (1973) Environmental and Birth Defects. Academic Press: New York-London, pp 305.

[6] Goncharov, N.V., Radilov, A.S., Mindukshev, I.V.,Kuznetsov, S.V., Ermolaeyeva, E.E., Glashkina, L.M., Kuznetsov, A.V. (2003) Pathophysiological Effects of RVX Low-dose Chronic Exposure in Experiments on rats. Ekonomika, logistika a ekolgie v ozbrojenych silach III. Sbornik z mezinarodni vedecka conference. BRNO, pp 63-67.

[7] Ermolaeyeva, E.E., Dobrylko, I.A., Goncharov, N.V., Mindukshev, I.V., Radilov, A.S. (2003) Studying the Kinetics of Thrombocyte Aggregation in Toxicological Experiment. Reception and Intracellular Signalization. Papers of International Conference. Pushchino, pp 63-65.

HEALTH AND ENVIRONMENTAL MONITORING AT THE AREA OF PROTECTIVE MEASURES IN THE VICINITY OF LOCATING CHEMICAL WEAPONS STORAGE AND DESTRUCTION FACILITIES

Sergey Nagorny[1], Andrey Radilov[1], Elena Tsibulskaya[1], Elena Ermolayeva[1], Fyodor Tsimbal[1]

[1]*Research Institute of Hygiene, Occupational Pathology and Human Ecology St.Petersburg (Russia)*

Abstract: The health and sanitary monitoring (HSM) in the vicinity of extremely hazardous destruction and storage facilities, and first of all at the territory where protective measures are conducted, should be of a specific type. Prompt and adequate solution of urgent problems of public health protection in the system of health and sanitary monitoring is achieved by developing the multi-level automated software-hardware unit on the basis of individual data bases and of monitored human environment factors. The systems and methods under consideration are now introduced at the area of the chemical weapons destruction facility in the settlement of Shchuch'ye in the Kurgan region.

Keywords: chemical weapons, destruction and storage facilities, sanitary monitoring.

The increased risk for human health in the vicinity of extremely hazardous chemical weapons destruction and storage facilities dictates the necessity of developing multilevel automated fast-response systems of supervising and monitoring the state of health and environment of the population residing in these areas. These systems should be developed on the basis of individual data bases and of monitored human environment factors.

The health and sanitary monitoring (HSM) in the vicinity of extremely hazardous destruction and storage facilities, and first of all at the territory of conducting protective measures, should be of a specific type. This means the necessity of monitoring chemical warfare agents (CWA) and their destruction products in a human environment, as well as taking into account specific syndromes and states observed during screening examinations of population and medico-statistical analysis of health indices typical for acute and chronic exposure to CWA. The system of special-purpose health and sanitary monitoring when conducting protective measures in the vicinity of chemical weapons destruction and storage facilities allows the problem of human environment safety to be conclusively solved during CWA storage and destruction.

V.M. Kolodkin and W. Ruck (eds.),
Ecological Risks Associated with the Destruction of Chemical Weapons, 315–319.

Prompt and adequate solution of urgent problems of public health protection in the system of health and sanitary monitoring is achieved by developing the multi-level automated software-hardware unit on the basis of individual data bases and of human environment factors monitored.

This unit includes the expandable and distributed information and analytical system (IAS) built through the architecture of "Client/Server." The system is intended to store, communicate, process, and analyze information that is contained in special data bases about the state of health, physiological and social status, harmful factors of labor conditions and environment of persons working at harmful industries, as well as of population of neighboring areas.

Developing the multi-level software-hardware unit includes three stages. The first -scientific and methodical stage - provides for creating and perfecting the data base structure and the whole scientific program on comprehensive monitoring. The second - organization and technical stage - provides development of the structure of the software-hardware unit including operating stations, commutation equipment, and communication support. The third - operational stage - provides for industrial operation of the system and correction of its operation in accordance with qualitative and quantitative changes occurring in the system "environment - population health."

The efficient interaction of the multiple components of the automated system is defined by the aims and tasks of all structural units. The basic structural unit of the automated system of the special-purpose health and sanitary monitoring is the information and analytical center (IAC), the tasks of which include forming, systematizing, processing, storing, and primary analyzing of the data base on the health state and environment of the population residing in the area where protective measures are conducted, in the vicinity of chemical weapons destruction and storage facilities.

The system analysis of data is carried out with the participation of specialists in the field of public health, toxicology, epidemiology, and occupational pathology working at the specialized Institute of the Federal Department "Medbioextrem" attached to the public Health Ministry of Russian Federation. At this level the interrelations "population health – human environment" are evaluated and causal relationships between operation of the chemical weapons destruction facility and its risk factors are determined. In this case these factors are differentiated from other causes of the environment associated diseases induced by industrial and natural factors. Besides, recommendations are worked out to perform sanitary and antiepidemic measures aimed at care and improvement of population health, and risk factors are revealed and predicted for population health and environmental state (see the Diagram).

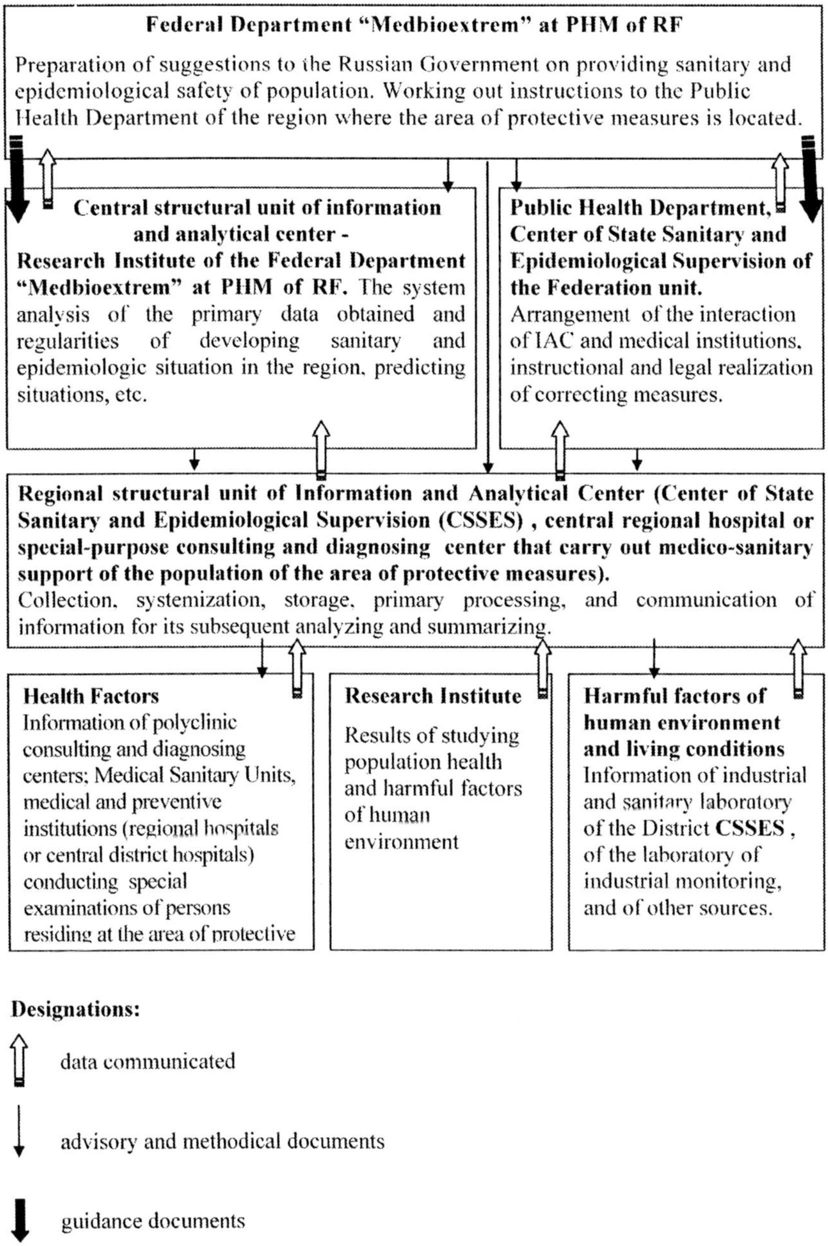

Figure 1: Interaction of structural components of the automated multi-level unit that affords conducting of health and sanitary monitoring at the area of chemical weapons destruction and storage facilities.

The Federal Department "Medbioexterem" attached to the Public Health Ministry of Russian Federation, Centers of State Sanitary and Epidemiological Supervision, and Public Health Departments of the regions where CW destruction and storage facilities are located carry out procedure and instruction functions with the aim of improving population medical support.

The distinguishing features of the special-purpose health and sanitary monitoring conducted in the areas where protective measures are conducted, in the vicinity of chemical weapons destruction and storage facilities, are as follows: high social implication, high legal responsibility, voluntary basis of examining persons, dynamism and functionality of the system, and the complexity of approaches in determining the association between the disease and functioning of chemical weapons destruction and storage facilities.

Effects of chemical burden on population health, ecologically responsible causes of its changes, as well as expert examination of the association of public diseases with exposure to harmful and hazardous factors of the environment, and the degree of "intensity" of the medico-ecological situation are evaluated in accordance with normative and methodical documents and recommendations [1-5].

The systems and methods under consideration are now introduced at the area of the chemical weapons destruction facility in the settlement of Shchuch'ye in the Kurgan region.

References:

[1] "Comprehensive sanitary evaluation of the intensity of the medico-ecological situation of various territories resulted from pollution of the population environment with toxicants." Methodical recommendations. Adopted by the State Committee of Sanitary and Eepidemiological Supervision of RF 30.07.97. Reg. N 2510/5716-97-32

[2] Maimulov V.G., Nagorny S.V. (2000) Methodical problems of studying casual relationships in the system "environment – human health". Journal St.Petersburg State Medical Academy named after I.I. Mechnikov. – SPb, 1, 22-35.

[3] Expert examination of determining the association between the population diseases and exposure of the body to harmful and hazardous factors of the environment. Manual for physicians. Adopted by the chairman of the public health section of the Scientific Medical Council Of Public Health Ministry of RF. 18.10, 2000, p 11

[4] Methodical recommendations for examining persons residing and working at the area of protective measures in the vicinity of chemical weapons destruction and storage facilities. Adopted by Federal Department "Medbioextrem" 15.11.2002, Moscow, 2002, p 56.

[5] Methodical recommendations for the course of conducting expert examination to determine the casual relationship between diseases of persons residing and working at the area of protective measures in the vicinity of chemical weapons destruction and storage facilities and these facilities functioning. Adopted by Federal Department "Medbioextrem" 22.05.2001, Moscow, 2001, p 23.

THE USE OF PUPILLOMETRY FOR EVALUATING THE FUNCTIONAL STATE OF PERSONS WHO WORKED WITH ORGANOPHOSPHORUS CHEMICAL WARFARE AGENTS

Fyodor Tsimbal[1]
[1]*Research Institute of Hygiene, Occupational Pathology and Human Ecology St.Petersburg, Russia*

Abstract: The use of the method of operative monitoring of the functional state of the workers by pupillometric findings to examine health of workers engaged in the process of chemical weapons destruction is suggested. To register pupillary responses and spontaneous changes of the pupillary diameter, the Special Design Engineering Office "Biophyspribor" in collaboration with the Research Institute of Hygiene, Occupational Pathology and Human Ecology has worked out several modifications of pupillometers built on modern components with the use of personal computers to process the obtained data in real time.

Key words: pupillometry, organophosphates, monitoring system.

The problem of chemical weapons destruction (CWD) involves the need for solving a whole complex of problems, among which the most important is the problem of the use of a monitoring system to examine the health of workers engaged in the process of CWD. The present paper suggests the method of operative monitoring of the functional state of examined persons by pupillometric findings as one of the components of such a system.

It is well known that embryogenesis of the visual organ is closely connected with brain embryogenesis. Considering this fact, and taking into account the peculiarities of structure, functions, and innervation of the visual organ, it can be regarded as a universal indicator of the functional state and diseases of the entire body.

By now, the industrial toxicology literature has accumulated a great number of data on changes in the visual sensor occurring after exposure of the body to harmful chemicals with various concentrations and periods of exposure. Such changes often appear earlier than somatic signs that are indicative of some changes in the body state, which is of practical use for the preclinical diagnosis of intoxication.

V.M. Kolodkin and W. Ruck (eds.).
Ecological Risks Associated with the Destruction of Chemical Weapons, 321–327.

One of the methods of investigating the visual sensor is pupillometry which is often treated as checking the pupil size in the examined. At the same time, thorough examination of all peculiarities and stages of pupillary responses to different kinds of stimuli and investigation of spontaneous changes of the pupillary diameter are of great diagnostic value. Such an approach elucidates a variety of informative parameters which allow us to describe adequately the processes occurring both in the visual sensor and in the human central nervous system as a whole.

To register pupillary responses and spontaneous changes of the pupillary diameter, the Special Design Engineering Office "Biophyspribor" in collaboration with the Research Institute of Hygiene, Occupational Pathology and Human Ecology has worked out several modifications of pupillometers built on modern components with the use of personal computers to process the obtained data in real time.

The pupillary response to the standard light stimulus (pupillogram) in healthy humans consists of the latent phase, the constriction phase, and the dilatation phase. It is customary to use the piecewise approximation to describe the pupillogram that allows some parameters of the pupillary response to be calculated. Among these are: D_i - the initial pupillary diameter (mm) prior to presenting the stimulus, t_l - the latent period of pupillary constriction (sec), t_c - pupillary constriction time (sec), t_d -pupillary dilatation time (sec), A_c - constriction amplitude (sec), V_c - constriction rate (mm/sec), V_d - dilatation rate (mm/sec). It is our opinion that "even" approximation (i.e., approximation with the function together with its derivative) of the pupillogram can be constructed according to the formula

$$D(t) = D_i - A^* \exp[-\alpha *(t-\tau)]*\mathrm{sh}[\beta *(t-\tau)]$$

The following conventional signs are used in the above expression: D(t) - pupillary diameter at the instant t, D_i – initial pupillary diameter with $t < \tau$, A - amplitude coefficient, α - attenuation constant, β- characteristic frequency, τ -time lag in the feedback circuit.

The above formula has been obtained in mathematical modelling of the pupillary reflex as a response of the system of regulating the intensity of retina illumination to a short light impulse. The parameters used in the formula are also pupillometric indices which are measured with the pupillometers described.

In healthy humans the pupillary diameter changes spontaneously in the course of time. This relationship can be processed by computers to obtain the frequency spectrum of the signal studied and to detect its quasi-periodic components. The data obtained are also of great importance for diagnosing the functional state of the examined.

The present pupillometric method was used for examining the diseased with chronic intoxication with organophosphorus chemical warfare agents (CWA) who have ceased working with these compounds 4-5 years ago (21 persons at the age of 30-50). The control group comprised 20 persons of similar sex and age.

Correlating the results of calculating parameters for both samples (see Table 1) allows the conclusion to be made that, in workers contacting with organophosphorus CWA, the time lag τ in the feedback circuit decreases significantly (Probability of the null hypothesis, $P < 0.01$), as well as the length of time τ_e in which the rate of changing the pupillary diameter reaches the extremum ($P < 0.05$), while the absolute values of this extremum (Ex) increase significantly ($P < 0.01$). It should be also noted that the parameter A in the same persons increases significantly ($P < 0.01$) compared to controls, α and β parameters drop significantly ($P < 0.01$). It was observed that in some persons with pupillary constriction who have worked with organophosphorus CWA (approximately in 30 – 50% of cases) non-montonous behavior of the relationship D(t) was registered at the trailing edge of the impulse, in many cases the picture of collision excitement of damping oscillations being noted. Such a tendency of changing parameters of the retina illumination regulating system is unequivocally indicative of the fact of turning the system into the overcontrol mode and hence of decreasing its stability. As a possible biochemical mechanism of decreasing the system stability, the mechanism of anticholinesterase effects of organophospohorus compounds can be suggested.

As the analysis of spontaneous changes of the pupillary diameter showed, they are certain to include the noise component and most commonly - one or several quasi-harmonic components.

During accumulating and analyzing information it was noted that in workers in contact with organophosphorus CWA quasi-harmonic components are found more often and they have a greater amplitude than the same components in controls. In some cases spontaneous changes of the pupillary diameters acquire a clear periodicity visible to the naked eye.

Tables 2 and 3 present the comparative data of the relative power of quasi-harmonic components of the pupillary diameter spontaneous changes in different groups stratified by certain signs. The groups indicated in Tables 2 and 3 were compared with the control group, consisting of persons who had no contact with organophosphorus CWA. The control group had the following indices: N = 20, $\gamma = 3.01$, m = 1.14. Table 2 gives the comparative characteristics by γ parameter in groups of persons who contacted with organophosphorus CWA, and in this case the groups were stratified by occupation. This division is consistent with the differentiation by the degree of contact of the personnel with harmful chemical factors. Laboratory workers working without self-protection means have essentially the greatest contact with organophosphorus CWA, instrumentation and control (I&C) men have slightly less contact and persons combined into the group "The rest" have the least contact with these chemicals. Comparing the values of γ parameter in the above groups shows that its value correlates with the extent of contact with organophosphorus CWA and decreases from 24.8% for laboratory workers down to 14.9% for persons of "The rest" group. The value of "t" criterion for the same groups decreases from 5.36 to 2.01, which correlates with the probability of the null hypothesis in the first case $P < 0.001$ and in the second case $P < 0.05$.

Thus, there is rather high significance in the first case as to the availability of statistically significant differences, while there are no such differences in the second case. For the group "Instrumentation and control men" there is statistically significant difference with the control group ($P < 0.001$) as well, at the same time the value of "t" criterion is somewhat lower for this group than for the group "Laboratory workers". Table 3 shows the comparative characteristics of γ parameter for the group stratified by the severity of the occupational disease. As can be seen from the given Table, in this case γ parameter also indicates the degree of damage to the human body with a harmful chemical factor. Both in the case of the early stage of the occupational disease and in its chronic course there is a significant difference by the given parameter with the control group ($P < 0.001$). It should be noted that γ value is slightly higher in chronic diseases which agrees with the mechanism of action of organophosphorus compounds on the human central nervous system.

The aspects of the early diagnosis of acute intoxication with the aid of pupillometry have been studied in the example of clinical trial changes to the functional state of examined, by exposing them to certain pharmacological drugs and subsequent elucidation of changing parameters of the pupillary response typical for these drugs. The particular types of pharmacological loads were chosen with the aim of modelling actual situations at chemical industries. Administration of proserine simulated the light degree of intoxication with organophosphorus compounds due to its anticholinesterase effects, while use of homatropine was consistent with the situation of simulation of overdosing the antidote, etc.

The investigations were carried out with the sample of 35 persons (14 men and 21 women) between the ages of 20-55. The results of measuring parameters of the pupillary response prior to and after the use of pharmacological drugs are shown in Table 4. Close inspection of these data shows that, for each pharmacological drug, there are its own typical changes of pupillary response parameters, which allows significant differentiation of the functional state of the examined to be carried out. It should be noted that, in these investigations, minimal quantities of drugs were used and in this case clinical manifestations were not reliably registered.

Thus, the findings of these investigations show that pupillometry can be recommended as a method for prompt revelation of intoxication in the personnel in contact with harmful chemical factors, organophosphorus CWA in particular.

Table 1: Parameters of the pupillary response model in workers who contacted with organophosphorus compounds and in controls

Model parameters	Persons who worked with organophosphorus compounds		Control group		t	P
	X	m	X	m		
1	2	3	4	5	6	7
Di	5.62	0.255	5.36	0.418	0.548	>0.05
Ex	4.14	0.338	2.768	0.238	2.69	<0.05
τ	0.369	0.023	0.58	0.053	3.73	<0.01
1	2	3	4	5	6	7
t_e	0.309	0.015	0.402	0.023	2.87	<0.05
A	3.339	0.142	2.276	0.260	3.87	<0.01
α	2.249	0.117	3.020	0.149	3.72	<0.01
β	1.65	0.137	2.796	0.193	4.60	<0.001

Table 2: Comparative characteristics of the relative power of quasi-harmonic components of spontaneous changes of pupillary diameters in different groups.
Stratification by professions.

Indices	Group			
	Laboratory workers	Instrument control men	The rest	Control
N	10	18	7	20
γ	24.80	21.56	14.93	3.01
m	5.36	4.81	9.85	1.14
k	28	36	25	---
t	5.28	3.93	2.01	---
P	<0.001	<0.001	<0.05	---

List of conventional signs:
N number of the examined in each group,
γ average value (for each group) of relative power of quasi-harmonic components of spontaneous changes of the pupillary diameter,
m standard deviation,
k number of degrees of freedom,
t Student's t-criterion,
P probability of the null hypothesis.

Table 3: Comparative characteristics of the relative power of quasi-harmonic components of spontaneous changes of pupillary diameters in different groups. Stratification by the occupational disease stages. (List of conventional signs as in Table 2.)

Indices	Group		
	Early stage of disease	Chronic disease	Control
N	13	9	20
γ	22.42	25.42	3.01
m	6.15	7.39	1.14
k	31	27	---
t	3.79	4.33	---
P	<0.001	<0.001	---

Table 4: Pupillary response parameters prior to and after certain pharmacological tests

Parametrs	Initial state		After injecting proserine		After injecting homatropine		After injecting pilocarpine		After alcohol use	
	X	m	X	m	X	m	X	m	X	m
D_i , mm	4.901	0.098	4.354***	0.086	6.773***	0.103	2.814***	0.130	4.777	0.153
A_c , mm	1.097	0.029	1.160	0.063	0.360***	0.043	0.453***	0.045	1.170	0.063
t_l , s	0.292	0.038	0.286	0.006	0.327	0.027	0.353***	0.012	0.279	0.0083
t_c , s	0.467	0.086	0.437	0.072	0.396	0.106	0.390***	0.021	0.480	0.013
t_d , s	1.765	0.044	1.959	0.142	0.686***	0.123	1.367**	0.122	1.712	0.098
V_c, mm/s	2.330	0.043	2.436	0.136	1.118***	0.169	1.170***	0.091	2.301	0.127
V_d, mm/s	0.573	0.020	0.554	0.102	0.507	0.090	0.327***	0.022	0.672*	0.045

Notes: *) $P < 0.05$; **) $P < 0.01$; ***) $P < 0.001$

5

SESSION IV: PUBLIC OUTREACH

PARTICIPATION OF THE PUBLIC IN LICENSING PROCEDURES FOR AND DURING OPERATION OF HAZARDOUS FACILITIES

Roland Fendler[1]
[1] *Umweltbundesamt (Federal Enironment Agency), Germany*

Abstract: The German legal framework includes several obligations on risk communication, which are relevant for hazardous facilities. These obligations take into account the requirements of business secrets, technical secrets and public security. Beside these obligations some operators experienced in the use of additional instruments for risk communication. All these instruments could be used in other countries as well, but should be modified according their own cultural risk communication background.

Key words: risk communication, environmental information, public private partnerships, hazardous facilities, business secrets, technical secrets, public security

1. ASPECTS OF PARTICIPATION OF THE PUBLIC

The free access to information and the rights of the public to participate in decision processes are important merits of a democratic society. Four main aspects of public participation are of main relevance in relation to the construction and operation of hazardous facilities:

1. The information of the public on risks due to the function of the public in political decision processes;
2. The participation of the public in administrative decision processes as a source for relevant information;
3. The participation of citizens in administrative decision processes due to the possible impact on their rights by the decision;
4. The information of the public on risks to allow appropriate behavior in case of emergencies to reduce the related consequences.

Nevertheless the rights of the public may be in conflict with other relevant interests like:

a) Business secrets;
b) Technical secrets;
c) Secrets due to the requirements of public security.

V.M. Kolodkin and W. Ruck (eds.),
Ecological Risks Associated with the Destruction of Chemical Weapons, 331–338.

In Germany, several laws and regulations are in application to ensure the free access to information and the participation of the public on one hand and guarantee the three kinds of secrets on the other hand. Beside these legal requirements, some authorities and some enterprises practice risk communication to the public by additional instruments.

"Public" here, is understood as the general population. It must to be borne in mind that risk communication is also required with the personal employed on the site of a hazardous establishment, respecting the special rights of, and the special risks for, personnel.

2. LEGAL FRAMEWORK IN GERMANY

For hazardous facilities three components of the legal framework are of main relevance

1. The Environmental Information Act;
2. The Federal Emission Control Act;
3. The Hazard incidents' Ordinance.

2.1 The Environmental Information Law

The Environmental Information Act is relevant for public authorities and gives the most general obligations in this subject. Environmental Information is all information in written, visual, aural or data-base form

1. on the state of the environment and
2. on activities or measures adversely affecting, or likely so to affect it, and
3. on activities or measures designed to protect the environment.

Public authority is any administration at national, regional or local level with public responsibilities and private or legal persons in charge due to them, and possessing information, relating to the environment with the exception of bodies acting in a judicial or legislative capacity.

These public authorities and natural or legal persons are required to make available all relevant information to any natural or legal person at his request and without his having to prove an interest. The access requires an application defining clearly which information is requested. If the request is legal the authority has two months' time to make the information available. The authority can charge a fee for the access to the information but this fee has to be limited.

A request for information can be refused where it affects:

- the confidentiality of the proceedings of public authorities, international relations and national defense,

- public security,
- matters which are, or have been, sub judice, or under enquiry, or which are the subject of preliminary investigation proceedings,
- commercial and industrial confidentiality, including intellectual property,
- the confidentiality of personal data and/or files,
- material supplied by a third party without that party being under a legal obligation to do so,
- material, the disclosure of which would make it more likely that the environment to which such material related would be damaged.

The Environmental Information Act has to be amended according to the new EU Directive of 28 January 2003 on public access to environmental information. This amendment may consider the UNECE Convention on Access to Information, Public Participation in Decision-Making and Access to Justice in Environmental Matters, called Aarhus-Convention; of 25 June 1998.

2.2 The Federal Emission Control Act

The Federal Emission Control Act and the related ordinances include i.a. the obligations for the licensing many different hazardous installations. A license by the competent authority is required for the construction, operation and any significant modification of a relevant installation. In Germany these licenses include other relevant licenses or permits, e.g. the construction permit and the licensing procedure may include an environmental impact assessment.

The applications documents of the operator have to be made available to the public for one month. Every one is allowed to raise objections in writing within one and a half month . The application and the objections have to be discussed. This discussion is organized by the competent authority and open to the public. The competent authority has to send the minutes of this discussion to the persons who made objections if they ask to do so. The decision on the license has to include the reasons for the considerations or refuse of the objections. The decision and the reasons for it have to made available to the persons who made objections or the public.

If the application documents include business or technical secrets the operator can apply to the competent authority to keep them confidential. The competent authority has to check this application for any information claimed as secret. If the competent authority agrees the operator has to provide to the competent authority a modified version of the relevant parts of the application documents for presentation to the public.

2.3 The Hazardous Incidents' Ordinance

The Hazardous Incidents' Ordinance ("Störfall-Verordnung") is the twelfth ordinance of the Federal Emission Control Act and the main instrument for German implementation of the European "Seveso-Directive". This ordinance is relevant for plants where large quantities of hazardous substances can be present on site. There are two tiers for these quantities according to the properties of these substances or for listed substances, giving "lower-level plant" and "upper-level plant". An plant can be one installation or a group of installations on a site controlled by one operator.

Four relevant instruments of the ordinance are:

a) The obligation of operators of "upper-level plant" to produce a safety report;
b) The obligation of operators of "upper-level plant" to draw up internal emergency plans and to supply information to the competent authorities, to enable the latter to draw up external emergency plans;
c) The obligation for authorities to inspect the relevant establishments and to prepare an inspection report on each inspection;
d) The obligation of operators of "upper-level plant" to inform persons liable to be affected by a major accident on the safety measures and on the requisite behavior in the event of an accident.

The safety report has to include information on the establishment, the related hazards, including possible consequences of major accidents, the safety measures that will be technical and organizational. Details on the content of a safety report are listed in Annex II and III of the ordinance and shall be defined by the administrative ordinance according to the Hazardous Incidents' Ordinance, which is under discussion. The operator has to review this safety report at least every five years and if a significant modification of the establishments requires this. The safety report has to be sent to the competent authority, which has to examine this report, and has to communicate its conclusions to the operator. The safety report or parts of the safety report have to be a part of the application documents presented to the public for the licensing of an installation according to the obligations of the Federal Emission Control Act. Furthermore the operator has to make the safety report available to the public.

The internal emergency plans must be drawn up by the operators in consultation with personnel employed in the plant and the authorities have to consult the public on external emergency plans.

The competent authorities have to organize a system of inspections of establishment concerned. Such inspections or other control measures must be sufficient for a planned and systematic examination of the systems being employed at the establishment, whether of a technical, organizational or managerial nature. Following each inspection, the competent authority shall prepare an inspection report.

Operators of "upper-level plant" have to supply information on safety measures and on the requisite behavior in the event of an accident, without their having to request it, to persons liable to be affected by a major accident. The information must be reviewed every three years and, where necessary, repeated and updated, at least if there is any significant modification of the establishment. The information must also be made permanently available to the public. This information has to include:

- An explanation in simple terms of the activity or activities undertaken at the establishment.
- The common names or, in the case of dangerous substances covered the generic names or the general danger classification of the substances and preparations involved at the establishment which could give rise to a major accident, with an indication of their principal dangerous characteristics.
- General information relating to the nature of the major-accident hazards, including their potential effects on the population and the environment.
- Adequate information on how the population concerned will be warned and kept informed in the event of a major accident.
- Adequate information on the actions the population concerned should take, and on the behavior they should adopt, in the event of a major accident.
- Confirmation that the operator is required to make adequate arrangements on site, in particular liaison with the emergency services, to deal with major accidents and to minimize their effects.
- A reference to the external emergency plan drawn up to cope with any off-site effects from an accident. This should include advice to cooperate with any instructions or requests from the emergency services at the time of an accident.
- Details of where further relevant information can be obtained.

If the safety report includes business or technical secrets the procedure is identical as for the licensing procedure, which means the operator, can apply at the competent authority to keep these information confidential. The competent authority has to check this application for any information claimed as secret. If the competent authority agrees, the operator has to provide to the competent authority with a modified version of the relevant parts of the safety report for presentation to the public or present the safety report with modifications to the public by its own.

If the safety report includes information with relevance for public security, the competent authority has to identify this information and instruct the operator to remove this information from the safety report presented to the public by the operator or has to remove this information if the authority makes the safety report or parts of the report available to the public.

The same procedures should be relevant for inspection reports, external emergency plans and information provided to the public by the operator.

3. ADDITIONAL APPROACHES

Some enterprises made the experience that these legal obligations are not sufficient for a well risk communication to the public. A favored additional instrument for risk communication is a communication forum. The enterprises organize these fora and invite authorities, local politicians and interest groups to attend. In these for a, the enterprises can present intended projects and activities, give information on the effects of the normal operation and even discuss causes and consequences in case of accidents. The politician's and interest groups can ask questions, demand information and propose new activities.

These fora make use of the experience that there are four groups in the public relevant for risk communication:

a) A large group which is not interested in any kind of information;
b) A smaller group which is interested in some information on risks;
c) A even smaller group which is interested in information on risks and asks for more information on the basis of the information provided to them;
d) Few "leading persons", in several cases very well educated and motivated, asking for detailed and actual information and having influence on the opinion of group c), b) and local and even regional politicians.

For group a) and b) the mandatory provision of information according to the Hazardous Incidents' Ordinance may be sufficient. A lot of enterprises publish additional information by the internet, which should be prepared according to the information demands of the group c). These two approaches are risk information. Some enterprises invite the "leading persons" according to group d) to their information fora and try to use them as a transmitter for their messages. Beside that the fora can allow the management of the enterprises to learn about the risk perception of the public from these "leading persons". Therefore the fora allow real risk communication and allow to communicate on risk problems before they may turn out into conflicts.

4. CONCLUSIONS

The German Environmental Information Act gives general access to the public to environmental information at authorities. This guarantees that the public is able to argue in political decision processes.

The participation of the public in licensing procedures, like according to the Federal Impact Control Act, allows the public to input information in administrative decisions and secures their rights that may be affected by these decisions.

The obligation of operators of "upper-level plants" according to the Hazardous Incidents' Ordinance, to inform the public on risks and appropriate behavior in case of accidents is an important instrument of related risk communication and shall reduce the consequences in case of emergencies.

All these legal obligations consider the required confidentiality of business and technical secrets and the requirements of public security.

Beside these legal requirements some operators use additional instruments for risk communication especially risk communication fora organized by them. According to their experience a forum is a good instrument to communicate with the public by "leading persons" of local interest groups.

These instruments of risk communication could be used in other countries as well but should be modified according to the differences in the cultural background of risk communication and even the history of risk communication relations.

NOTES

More advice on industrial plant safety and risk communication is included in the OECD-Guiding Principles for Chemical Accident Prevention, Preparedness and Response (2nd Edition 2003) http://www1.oecd.org/scripts/ehs/guidingprinciples/index.asp

References

[1] Council Directive 90/313/EEC of 7 June 1990 on the freedom of access to information on the environment Official Journal L 158, 23/06/1990 P. 0056 – 0058.

[2] Directive 2003/4/EC of the European Parliament and of the Council of 28 January 2003 on public access to environmental information and repealing Council Directive 90/313/EEC; Official Journal L 041, 14/02/2003 P. 0026 - 0032 http://europa.eu.int/eur-lex/pri/en/oj/dat/2003/l_041/l_04120030214en00260032.pdf.

[3] UNECE Convention on Access to Information, Public Participation in Decision-Making and Access to Justice in Environmental Matters, called Aarhus-Convention; of 25 June 1998. http://www.unece.org/env/pp/documents/cep43e.pdf.

[4] Council Directive 96/82/EC of 9 December 1996 on the control of major-accident hazards involving dangerous substances; Official Journal L 010, 14/01/1997 P. 0013 - 0033; http://europa.eu.int/smartapi/cgi/sga_doc?smartapi!celexapi!prod!CELEXnumdoc&lg=en&numdoc=31996L0082&model=guichett.

Printed in the United Kingdom
by Lightning Source UK Ltd.
109292UKS00006BB/23-24